2/15

EMERGENCY MANAGEMENT
and
SOCIAL INTELLIGENCE

A Comprehensive All-Hazards Approach

EMERGENCY MANAGEMENT
and
SOCIAL INTELLIGENCE

A Comprehensive All-Hazards Approach

Charna R. Epstein
Ameya Pawar
Scott C. Simon

CRC Press
Taylor & Francis Group
Boca Raton London New York

CRC Press is an imprint of the
Taylor & Francis Group, an **informa** business

CRC Press
Taylor & Francis Group
6000 Broken Sound Parkway NW, Suite 300
Boca Raton, FL 33487-2742

© 2015 by Taylor & Francis Group, LLC
CRC Press is an imprint of Taylor & Francis Group, an Informa business

No claim to original U.S. Government works

Printed on acid-free paper
Version Date: 20141024

International Standard Book Number-13: 978-1-4398-4797-8 (Hardback)

Visit the Taylor & Francis Web site at
http://www.taylorandfrancis.com

and the CRC Press Web site at
http://www.crcpress.com

CONTENTS

SECTION II   *Social Intelligence Framework and Intelligence Methodology and Emergency Management*

SECTION III   *Program and Policy Prescriptions*

INTRODUCTION

For a moment, put aside everything you know about emergency management. Forget the plans, the go-kits, and the first responders. Now think about the Central Intelligence Agency (CIA) and the work it does. What words come to mind when you think about intelligence gathering and analysis: spy, espionage, surveillance? Let's think beyond the cloak-and-dagger clichés of intelligence agencies and think about the type of work they really do. Policy makers depend on information to make decisions. The U.S. president, Congress, and military leaders depend on intelligence agencies to provide information on nations in order to make informed decisions. If we remove the espionage and intrigue, intelligence agencies spend a bulk of their time trying to make sense of what is happening in a foreign country and reporting this to key decision makers. This means they are trying to understand the country by studying the people, the culture, the economy, etc. In sum, they are providing a cold analysis of a situation. The intelligence analyst's job is to remove his or her politics from the equation and provide analysis to a decision maker. Why? Foreign policy decisions are made with U.S. interests in mind in the near term and long term. By garnering intelligence, as much intelligence as is available, the hope is to make more informed decisions. Informed decisions reflect an intelligent information gathering and decision-making process. All said, the decision maker is free to insert his or her politics, or the politics of the day, into any decision that is made, but the starting point is rooted in a matter-of-fact intelligence analysis.

Did you know during a disaster human beings were used as sandbags to prevent floodwaters from entering certain areas of a city? Or did you know there are communities where a historical mistrust of government makes emergency evacuation orders impossible to enforce? These are all historical issues that came up during emergency response and recovery operations in the United States. These issues also constitute a type of intelligence—social intelligence. Knowing and understanding historical issues relative to disaster events is important, and having some understanding can prevent a small problem from metastasizing into a catastrophic mistake. In the intelligence world, understanding tribal rivalries in Afghanistan or making sense of separatist groups in Georgia can be the difference between life and death. Yet, applying these international frameworks domestically in order to save lives has never been contemplated.

Another important set of factors to consider is how intelligence is utilized by any national security apparatus. Academics and think tanks work on conceptual issues, study long-term strategic issues, and provide subject matter expertise to policy makers. Intelligence agencies aggregate on-the-ground intelligence, military intelligence, publicly available information (news sources), with applicable information from sources and covert operations to develop an understanding of what is happening in a region or country. When all of this information is cobbled together, policy makers make decisions based on this information. Policy makers insert their politics or the politics of the moment into their decision, but the intelligence-making process is largely devoid of prevailing political winds. That is, those tasked with providing intelligence provide a cold and analytical view of the region or country they study.

Wouldn't emergency management be better served if they understood the historical and real-time issues impacting communities prior to disaster? Of course—enter social intelligence.

To begin, think about all the post-disaster press conferences and think about what politicians and policy makers say. Time and time again, the conversation post-disaster revolves around people not doing enough to take care of themselves or their families. Little consideration is given as to *why* people can't prepare, respond, or recover from disasters. This lack of consideration creates an enormous gap. Emergency management and government do not fully understand the people they serve.

Did you know that droughts are the most expensive and regularly occurring disasters impacting our society? Or did you know that poverty and vulnerability aren't static states of being? Finally, did you know that there are communities within communities and, in addition, that the connections between people and their communities are what actually drive what happens during and after disaster?

Intelligence agencies ask these types of questions in a security-related context as they inform policy makers about what is happening in a country. They do so because policy makers try to understand how a decision they make can have a ripple effect in the target country or region. Intelligence agencies try to provide as much context as possible prior to making a decision. Unfortunately, when it comes to domestic policy, and especially as it relates to emergency management, there is little attention paid to what is actually happening in communities predisaster.

How does any of this apply to emergency management? It's simple—state and local governments have done little to fully understand how communities ebb and flow. Making sense of preexisting conditions requires gathering intelligence related to local economies, populations, tax collections, and hundreds of other qualitative and quantitative data points. This information is pieced together to provide a full picture of what is happening in a community. The result: social intelligence. This information on preexisting conditions directly impacts how communities and people prepare, respond, and recover from disaster. Understanding these preexisting conditions can also result in better emergency management systems and can allow government to avoid response and recovery failures. Furthermore, in disaster recovery efforts, this information can be used to bring communities *back* to preexisting conditions, or better optimally, for improved functioning and resiliency.

Equally important is recognizing the connections between the disaster event and the more global impacts beyond the localized impacts; that is, understanding how emergency management addresses one disaster impacts what happens regionally, nationally, and even internationally. Clearly, there are lessons learned from disaster events during response and recovery efforts. The question is: Are we using this information to ensure better results the next time around?

We would argue that we collectively are not. Because if we were, we would not have written this book. So to better harness lessons learned and to avoid repeating the same mistakes over and over, we began researching disaster response and recovery and the underlying philosophy of emergency management.

Having an emergency plan for your community is different from being truly prepared. Hurricane Katrina, Sandy, and recent super-storms point to a frightening reality—these disasters are occurring more frequently, while emergency management systems across the country have done little to understand why disasters do not impact everyone equally. In order for a community or organization to be truly prepared, emergency management systems must begin to understand how a disaster event impacts physical infrastructure and people. This means that for a given community, emergency management must take into account how people interact with one another, how they interact with government, and how they deal with the disaster event. In order to understand these complex interactions, emergency managers must understand the populations they serve prior to disaster. This book will provide emergency managers, policy makers, and elected officials with a comprehensive framework for how to understand their community before, during, and post-disaster.

SUPER-STORMS AND 100-YEAR EVENTS

Climate change is real—each season brings more dramatic and severe weather patterns. In fact, we are seeing communities experiencing 100-year flood events every few years. Tropical storms are bigger. Tornadoes are more frequent, deadly, and destructive. Hurricanes and typhoons are more ferocious. Snow events and blizzards are shutting down cities for days and weeks. This is the new normal. As the climate changes, so must communities. This means understanding that areas once considered ripe for farming may experience more drought conditions. Regions once considered unsuitable for farming may present opportunities to raise certain types of crops. And cities like Chicago, Detroit, and Minneapolis, and others with access to an abundance of fresh water, must adapt to a future where they will see massive population growth because extended droughts will force people to locate near fresh water. While these global changes occur, communities must begin to understand how a changing climate impacts their daily lives. For a policy maker or elected official, climate change issues can no longer be kicked to a higher level of government. Emergency managers know all response is local. Without localized thinking on climate change, disaster impacts will continue to intensify.

In fact, in May 2014, the White House and President Obama released the third U.S. National Climate Assessment (NCA). The report details how the impacts of climate change can be felt throughout the country. Specifically, summers are longer and hotter. There are more frequent droughts. When it does rain, rainfall amounts are more extreme and flooding is a frequent issue communities are dealing with. During the winter, extreme cold is followed by very heavy snowfalls and followed again with extreme cold. Climate change is impacting every corner of the United States—the same is true around the world. Climate change impacts our daily lives prior to any disaster. The changes in our lives must be understood, and social intelligence can help emergency managers build systems that take the impacts of climate change into account as they prepare their communities for disaster events.

Beyond simply adapting to climate change, communities will have to fundamentally change disaster preparedness and response considerations. Disaster events linked to a changing climate are occurring more frequently, and these disasters are having a major impact on communities. For example, in many major urban areas, rapid development has led to a degradation of the amount of green and open space. In many cities, developers and buyers seeking to maximize floor space within a building traded outdoor and green space for living/commercial space. Over time, and when combined with climate change, the degradation of green and open space is leading to major flooding issues. Cities of all sizes have been slow to adapt to increasing rainfall and snow events. Localized flooding is a new norm, and aging sewer and storm water management systems will not be able to handle the amount and intensity of rain events. This means that communities may experience more drought and more severe rainstorms when there is rainfall. How does this play out locally?

In Cook County and Chicago, the Metropolitan Water Reclamation District (MWRD) invested nearly $3 billion into developing a deep tunnel system. This deep tunnel system project was started almost four decades ago and was developed so that Chicago and surrounding suburbs could better manage storm water. By managing storm water more effectively, rain events would not force the MWRD and City of Chicago to release raw sewage into the Chicago River and Lake Michigan.

> Rains of greater than 2.5 inches a day, the amount that can trigger sewage dumping into Lake Michigan, are expected to increase by 50 percent between now and 2039, according to a study by scientists from the University of Illinois at Urbana-Champaign and Texas Tech University. By the end of the century, the number of big storms could jump by a whopping 160 percent.[*]

This infrastructure project was one of the costliest in U.S. history, and for a long time was seen as a potential model for other cities. Then the climate began to change. The deep tunnel system was built to handle what were normal rain events. The new normal is one where rain is infrequent (outside the norm anticipated for the deep tunnel project), and more intense when there are rainfalls. Even with the added capacity, Chicago's storm water system is regularly overwhelmed and sewage is regularly released into Lake Michigan. The lake is also the source of drinking water for millions of Illinoisans, and residents of Indiana and Michigan. According to the National Resource Defense Council, the increased flooding will result in a range of health impacts and risks, including:

---

[*] http://articles.chicagotribune.com/2011-04-21/news/ct-met-deep-tunnel-climate-change-20110420_1_climate-change-sewers-deep-tunnel-project.

- Death and injury
- Contaminated drinking water
- Hazardous material spills
- Increased populations of disease-carrying insects and rodents
- Moldy houses
- Community disruption and displacement

The Chicago example is not unique. Along the coasts, hurricanes are expected to grow in size, complexity, and severity. Researchers from MIT and Princeton University have found that with climate change, such storms could make landfall far more frequently, causing powerful, devastating storm surges every 3 to 20 years. The researchers simulated tens of thousands of storms under different climate conditions, finding that today's 500-year floods could, with climate change, occur once every 25 to 240 years.*

A Hurricane Katrina-type storm will likely occur more often and may come in the form of a super-storm. In fact, nearly 10 years after Hurricane Katrina, a new, more powerful storm named Hurricane/Super-Storm Sandy devastated the Eastern Coast of the United States. This storm caused over $50 billion in damage and 72 people lost their lives.† The storm surges and wind gusts damaged thousands of structures, flooded subway lines in New York City, and rendered thousands homeless. These types of storms and disasters were once rare events. They are occurring more frequently and with greater intensity, and it is time emergency management begins to prepare for climate change-related disasters by evolving the field and taking more time to understand communities on the front end to better help them recover on the back end.

So what can emergency management do? Evolve, change, and most importantly, harness its preexisting strength to convene important conversations about where communities are today and where they might be after a disaster.

ECONOMIC REALITIES OF DISASTER RESPONSE

Not convinced that emergency management should evolve? Think about this. Since 1980, 144 weather-related disasters have resulted in over $1

---

* "Storm of the Century?" Try "Storm of the Decade," http://web.mit.edu/newsoffice/2012/storm-of-the-decade-0213.html.
† National Weather Service, http://www.nhc.noaa.gov/outreach/presentations/Sandy2012.pdf.

trillion in damages and costs. More importantly, we are no better prepared after having spent the money than prior to the disaster event.* A trillion dollars is a lot of money. We are not arguing that the entire dollar amount was wasted. We are saying that communities should be more resilient after having spent $1 trillion than prior to disaster. What does this mean? Communities evolve as the economy ebbs and flows. Communities change as people move in or out, demographics shift, and now, as the climate changes. These changes should be incorporated into how we prepare, respond, and recover from disaster. For example, after a major flood or a hurricane, communities think about hardening critical infrastructure and reexamining building codes. Policy makers should also focus on amending zoning codes to create more open and green space to allow for better storm water management for the next flood or hurricane. This means that in addition to focusing on creating resilient communities by making people more resilient, it is important for policy makers to build communities so they are ready for the future. And by doing so, they will prepare their communities, and the people they serve, for future disasters. The result: community resiliency that encompasses government, infrastructure, and people.

EMERGENCY MANAGEMENT TODAY AND WHERE IT NEEDS TO GO

Traditional emergency management systems develop emergency plans based on a first responder model. That is, plans call for response and recovery actions that enable emergency managers to quickly distribute resources and subsequently turn over recovery activities to other organizations. Simply put, response and recovery actions are based on a one-size-fits-all response based on the responding agencies, size, resources, capabilities, and expertise; however, it is not always focused on the affected community's needs. This one-size-fits-all approach works when populations are resilient; however, recent disasters provide clear evidence that long-term recovery is, for most disasters, the new norm.

How does an emergency manager better plan for disaster? Emergency managers must acknowledge that disasters are classified as such when there is an impact to humans. For decades, emergency management has focused on hardening physical infrastructure and securing resources

* http://www.usatoday.com/story/weather/2013/06/13/weather-climate-disasters/2419957/

to deal with a disaster for a 72-hour response/recovery model. While successful in some cases, the U.S. continues to spend billions of dollars annually on long-term recovery. A more robust emergency management system should reflect the relationship and the interconnections between social systems and physical infrastructure. More importantly, emergency management must take into account how we view preexisting social problems that impact disaster and begin to develop plans and systems that reflect how communities ebb and flow. It is the role of emergency management to ensure the components are intelligent of preexisting conditions prior to disaster.

Understanding the preexisting conditions in a community is what we call social intelligence (SI). For a community to be socially intelligent, we call for the collection and aggregation of data related to preexisting conditions. Emergency management should lead the effort in gathering, analyzing, and disseminating the data in real time related to these conditions. In increasing amounts, the information is "out there" and available.

Prior to writing this book, we presented our concepts related to SI at conferences. At each presentation, we received pushback on SI—emergency managers have been resistant to incorporate real-time planning into their emergency management systems.

> An approach that strives to reduce social vulnerabilities as well as respond to them challenges emergency managers to move beyond obvious needs—for sign language interpretation, for example, or life-preserving medical equipment or services—to consider the less obvious potential of vulnerable populations as partners throughout the disaster cycle.*

We have gotten pushback regarding this idea, but really collecting this information is just a domestic application of what we already do in the international area as it relates to our national security. National security is a confluence of domestic priority merged with international interests. The national security act mandates that we look at domestic and international information to ensure comprehensive international security. But domestically we have done very little analysis on how social systems, economics, physical infrastructure, and other stimuli contribute to national health and security.

As we stated earlier, we estimate that nearly $1 trillion has been spent on disaster-related activities. Many of these dollars have been expended

---

* Enarson, Elaine, Identifying and Addressing Social Vulnerabilities, Chapter 13, *Emergency Management: Principles and Practice for Local Government.* December 1, 2007 by William L. Waugh, Jr. (Editor), Kathleen Tierney (Editor), ICMA Publ., Washington, DC.

on response and recovery operations, and we are no more prepared after spending those dollars than we were prior to these expenditures. Had we spent those dollars more efficiently, our communities, perhaps our nation, could have been made more resilient. Going forward we must realize the world is changing, information is more abundant, communication travels a lot quicker, the impacts of globalization are starting to take shape, and climate change will increase the frequency of disasters. These emerging trends demand a new way of doing things in emergency management (EM). We must be more efficient with our resources. To achieve this efficiency and increase intelligence within EM systems, we must leverage and model techniques that exist as part of our national security infrastructure to increase our domestic competence.

The national security infrastructure that we are referring to involves leveraging intelligence methodology, intelligence analysis knowledge, and intelligence information architecture. We don't need to reinvent the wheel. To increase domestic competence, we have developed a social intelligence framework. This conceptual framework will provide you with the necessary information to increase situational awareness.

After having studied various disasters and their outcomes, we saw trends. The same issues related to preexisting conditions come up after most disasters. Yet, emergency management has not found a way to address these preexisting conditions in a way to make communities more resilient for future disaster events. Many of these preexisting conditions are the result of social problems. While it is not in emergency management's purview to solve these problems, emergency management must nevertheless *deal* with these issues during a crisis, and the subsequent response and recovery. To deal with these issues in a more robust fashion, we believe we must leverage existing emergency management principles with our social intelligence conceptual framework. Social intelligence can be best defined as

> a method by which emergency management gains critical situation awareness of a community in relative real time. This is achieved through a process of leveraging existing data, information, social capital, and preexisting conditions to develop a composite understanding of what is happening in a community. Making sense of preexisting conditions requires gathering intelligence related to local economies, populations, tax collections, and hundreds of other qualitative and quantitative data points—this information is pieced together to provide a full picture of what is happening in a community.

Once a community has adopted this approach within its emergency management framework, the operational framework for SI can be defined by the following statement:

The process for acquiring social intelligence involves the collection, aggregation, and interpretation of quantitative baseline data points (BDPs) and the integration of those BDPs into a qualitative architecture. The integration of BDPs into a qualitative architecture enables the emergency manager or policy maker to understand the connections between upstream, downstream, and cross-stream dependencies. The integration also allows the emergency manager or policy maker to make more informed estimations and decisions for all-hazards planning regarding preparedness, mitigation, response, and recovery for a given community. SI is a system for efficient access of unique local data to create a view of community strengths and challenges as they relate to emergencies. This type of situational awareness is social intelligence.

Our SI model recognizes that information is critical—it can mean the difference between success and failure. To implement SI architecture, an emergency manager should follow these three steps:

- Identify baseline data points (BDPs) across municipal, NGO, and corporate entities
- Aggregate BDPs to gain jurisdictional competence and real-time situational awareness
- Leverage information to align with emergency management systems

The SI model requires the collection of BDPs within a jurisdiction. Your jurisdiction may cover a municipality, county, state, or region. The SI data points reside in the public, nonprofit, and private sectors. This information often goes unnoticed by emergency managers. It is these data points that can provide the emergency manager with a more robust understanding of your jurisdiction. After collecting this information, it should be aggregated. The aggregation of the information provides emergency managers with real-time situational awareness. This situational awareness can then be leveraged with emergency management systems and ensure that preparedness and response systems remain living. Increased situational awareness can also help inform emergency managers of resource needs, gaps, and other information critical during a response and recovery.

Many of the BDPs we detail later in the book are quantitative measurements. For example, some BDPs include number of beds in shelters, unemployment rates, infrastructure age, available fund balance, and municipal debt load. By collecting these quantitative data points, emergency managers can then make qualitative assessments. For example, emergency managers may be able to make connections between access to mass transit, poverty rates, and economic development. Or they may be able to understand the connections between infrastructure, capital improvements, and property tax collection percentages. While BDPs may not be emergency management focused, they play a direct role in how emergency plans are developed. The BDPs can also give an indication of how people may be feeling about the government, and how they may respond or subsequently recover. Our SI model will allow you to connect hard data against other hard data and provide the framework to connect those data with qualitative trends.

Later in this book, you will read about how technology can help an emergency manager or policy maker build a socially intelligent emergency management system. As state earlier, the BDPs currently exist in silos. Even after identifying the BDPs relevant to a community, it is important for a socially intelligent system to understand the connections across the various data points and volumes of information. Technology can help create a visual representation of information. While information technology is a key component of implementing our social intelligence model, it is just a value-add. In our experience, we've see organizations purchase technology tools in an attempt to create better emergency management systems without first developing a strategy around what they hope the technology will help them achieve. Ultimately, the technology is only as good as the inputs. What matters most is ensuring that emergency managers are looking at the right information, drawing information about a community from a variety of sources (BDPs), and using this information to ask more questions to understand their communities. Once this process is implemented, technology tools like geographic information systems (GISs) can provide spatial understanding of what and where things are happening in a community. Other platforms, like Open 311, can help communities better connect service requests to needed capital improvements, and in turn reduce vulnerabilities. Finally, we want to reinforce that technology is a tool to help implement social intelligence, and our model is not meant solely for large municipalities or organizations. We understand that for a small community implementing Open 311 is not possible—but we do believe that every community and every emergency manager can harness information that already exists. Later in the book you will learn more about how to leverage existing technology tools to implement social intelligence.

Ultimately, the collection, aggregation, and analysis of the BDPs create a SI emergency management system. The information collected in a SI system can be leveraged by upstream, cross-stream, and downstream partners to mitigate the very circumstances that stress emergency management systems. Some information may prove useless. You may also find that most of the information is not actionable. That is, you may collect and aggregate BDPs and then realize nothing can be done to mitigate or change a situation. That is okay—at least you will be aware of what you can and cannot do. The more information you have, the more informed your system.

Social intelligence is a domestic application of how the national security apparatus develops foreign policy and actions. While one component of the U.S. national security apparatus clandestinely gathers information and takes action, most of the agencies compile information and aggregate information (open source, academic, news, etc.) and combine it with data/information collected clandestinely to provide a picture of what is happening around the world. By collecting, aggregating, and analyzing data and information (both quantitative and qualitative) prior to a disaster, for the communities we serve, we can ensure response and recovery is tailored for the people we serve. This is a more efficient, cost-effective, and sound way forward. Moving away from one-size-fits-all planning templates and plans toward more efficient emergency management systems would be akin to how our national security apparatus is assembled. More specifically, the national security infrastructure that we are referring to involves leveraging intelligence methodology, intelligence analysis knowledge, and intelligence information architecture. To increase domestic competence we have developed a social intelligence framework. The social intelligence framework will provide you with the necessary information to create a more intelligent emergency management system.

In Section I of this book, you will learn from case studies that identify common issues that impact emergency management systems during and after disaster. Since 1980, 144 weather-related disasters have resulted in over $1 trillion in damages and costs. More importantly, we are no better prepared after having spent the money than prior to the disaster event.[*] After reading case studies, you will be provided with a comprehensive analysis of social welfare policy and emergency management policy. This section will provide you with a solid foundation of the mirroring effect emergency management and social welfare policies have had on

---

[*] http://www.usatoday.com/story/weather/2013/06/13/weather-climate-disasters/2419957/

one another. At the end of Section I, you should begin to understand why emergency managers must be more informed about their communities. This will lead you to Section II, which focuses on intelligence from a national security perspective. National security policy is about the collection, aggregation, and analysis of various pieces of information and data. This process makes the national security apparatus more intelligent and informs policy makers with more robust information prior to making decisions. There is a specific science and methodology to this intelligence process. Before you learn about how to make your emergency management system more intelligent, we will examine the U.S. intelligence community's history, its agencies, and various methodologies. From there, you will learn how intelligence agencies help policy makers identify actionable intelligence. Then you will learn about the impacts of intelligence on foreign policy decisions—you will read about the impacts of good and bad intelligence. And lastly, you will read about how intelligence agencies survey the geopolitical landscape to help policy makers make informed decisions related to national security and U.S. interests. In Section III, you will learn how to develop SI architecture for your jurisdiction, as well as various information sources you can tap into. But before you learn how to develop and implement SI within your emergency management system, you must understand why critical information related to emergency management has gone untapped.

We believe that emergency managers should collect data, aggregate information, and ultimately understand daily life in the communities they serve—this is key to developing socially intelligent emergency management plans and systems. While planning for communities that have high rates of poverty will certainly result in better preparedness, response, and recovery efforts, that argument is only a piece or our social intelligence thesis. What we are telling emergency managers, public servants, government officials, etc., is that by knowing your communities, knowing your populations, you can harness the inherent resiliencies that accompany specific communities. The Lower Ninth Ward might have high rates of poverty, but it also had deeply connected neighborhoods where people knew each other and depended on one another. In the parish system of New Orleans, you had communities tied to their churches. This gives a reliable, trustworthy, and competent mechanism for education and communication. These communities aren't vulnerable because they aren't resilient or are faulty in some way; they have vulnerabilities because they can't depend on or trust traditional forms of assistance, communication, and—in many cases—government. There is a disconnect between government and traditional response and recovery entities, not from resiliencies. Knowing a community can help harness specific resiliencies while helping government operate in a more efficient and intelligent manner.

To reiterate, vulnerability as we talk about it in this book is not inherent or innate to specific communities referred to as vulnerable. Vulnerability is solely a matter of being less connected to traditional government services; it's really more a statement on government and its inability to serve all people equally. Vulnerability is a state of being less easily served and less connected to government services, for many reasons. Identifying vulnerable populations is important because it helps us ask better questions about serving all communities. And, the good news is, being deemed vulnerable absolutely does not mean less capable. There are lots of communities that we refer to as vulnerable that are more resilient in many ways than more mainstream communities. That's why understanding our communities is important. It doesn't take a lot to turn a vulnerable community into a more resilient community by harnessing inherent resiliencies. For example, while a community might have a high rate of poverty, it might also have built-in parish systems, block clubs, and multigenerational households— all networks that create strength within a community. But if emergency managers do not know about inherent strengths within a community, they may just see weaknesses or vulnerability. Simply put, vulnerability can be mitigated by understanding and leveraging strengths. This requires intelligence—social intelligence, and this book will provide a comprehensive understanding of why social intelligence is necessary and how you can use the system to better prepare your community for disaster.

This book will do the following:

1. Provide a rationale for why issues related to social vulnerability are misunderstood, and provide the foundation for social intelligence.
2. Go through various disasters via case studies. You will see the same issues arise during and after each disaster.
3. Begin to unpack how perception and the socially constructed views of the vulnerable drive flawed disaster response and recovery policies.
4. Provide the rationale for leveraging the intelligence framework currently used by the national security apparatus for domestic use—social intelligence.

5. Identify various social intelligence baseline policies as well detail how to implement the system.
6. Detail how to use GIS technology to provide visual context to social intelligence.
7. Detail some prescriptive policies and programs using social intelligence to build more resilient communities.

Section I

Emergency Management and the Need for Social Intelligence

1

What We Have—
and Haven't Learned

The terrorist attacks of September 11, 2001, changed the way U.S. federal agencies and state and local authorities interfaced with one another: all levels of government were forced to work collaboratively. Additionally, after 9/11, the Department of Homeland Security was created so that the U.S. intelligence, disaster, border protection, and federal investigative agencies would work together and share information. Around the world, foreign intelligence agencies began to adapt to respond to a global war on terrorism. Intelligence agencies began to share information and coordinate law enforcement and intelligence operations. Why? Yet as nations worked collectively to respond to threats of terror, emergency management lagged. In the United States, the impacts of Hurricane Katrina were a good example of how emergency management lagged behind the innovations in intelligence. Emergency managers and policy makers failed to understand their population on the front end. So during the response, they were unable to understand why people ignored evacuation orders. Later in this book you will see many examples throughout history where emergency managers and policy makers failed to harness information to inform emergency management plans and operations.

The world watched as people frantically waved for help from rooftops, waded through sewage in search of food, and cried for help from the Superdome (Figure 1.1). The massive hurricanes hit the Gulf Coast with enough fury to breach the New Orleans levee systems. New Orleans, an iconic American city, was forced to its knees. Eight years later, much of New Orleans

Figure 1.1 New Orleans, Louisiana, August 31, 2005. Local residents arrive at ramp to the Superdome after being rescued from their homes. New Orleans is being evacuated due to flooding caused by Hurricane Katrina. (From Jocelyn Augustino/FEMA.)

looks no different than it did in the days and weeks after the water receded. Much of New Orleans has yet to be rebuilt, and thousands of her residents are scattered across the nation. There are still a few hundred families living in Federal Emergency Management Agency (FEMA) trailers. Even today there are many policy makers that still believe that a failure to follow the self-evacuation order was a primary reason for the failed response. A familiar quote also comes up frequently: "We told you to leave. Why didn't you leave?" The answer to that question is the responsibility of emergency management.

To answer that question, emergency management must move beyond template planning and develop a comprehensive understanding of their jurisdiction. Social intelligence is the aggregation of data points, qualitative and quantitative, that provide information regarding a population's access to or amount of jurisdictional capital. What we have learned since 2005 is that many people in New Orleans were unable to evacuate because they did not have access to a car. Many did not have access to credit or other financial instruments that could aid them during an evacuation. Some people simply did not have anywhere to go. Many others had cars but very little money and could not fill up their gas tank. Some people were hesitant to leave because they did not trust the government warnings, and many others had some combination of the factors listed above. Yet when the mandatory evacuation order was given, there was not an effort to get to these people. Instead, state and local governments pushed liability down to the individual. What state and local authorities failed to realize is that nearly 20% (100,000 of the 500,000) of the population of New Orleans was socially vulnerable to disaster. That is, they would not be able to prepare, self-evacuate, and subsequently recover without aid from the government. How could authorities not have known?

We believe the answer is simple. Social vulnerability and poverty are complex states of being. They are not static, and often vulnerability can vary from summer to spring, paycheck to paycheck, or even day to day. Vulnerability comes in many forms: the poor, the elderly, the unemployed, single-parent households, returning veterans, groups who lack professional portability, minority groups for whom English is a second language, etc. Making sense of a community is difficult for emergency management because it tries to build emergency plans based on static assumptions. Think about it this way: When a wildfire or mudslide forces an evacuation of Malibu, authorities can be certain that nearly all of the residents can self-evacuate. Authorities can also be certain that residents of Malibu can rely on insurance, personal savings, and other resources during recovery. Responding authorities are more concerned with life safety operations than with any social impacts. But

Malibu is not every American city. In fact, it is one of the few examples where we have homogeneity in terms of access to financial and social capital. So, a self-evacuation order would almost likely be followed 100% of the time. Understanding how different communities will react and respond is social intelligence. Emergency managers should know ahead of time which neighborhoods or communities can respond, at what level, and which cannot. Social intelligence also means understanding that even though a community could self-evacuate at the 100 percent rate 4 years ago, the same might not be possible today. A community can change quickly due to unemployment rates, immigration, or a local budget crisis. This means a community that was highly resilient a few years ago might no longer be able to respond and recover similarly due to a changed socioeconomic state.

Imagine you are the emergency manager for Los Angeles County. A regional event impacts the county that requires self-evacuation. You can be sure that the residents of Malibu can evacuate, but what do you do about Watts? Watts is a neighborhood of Los Angeles where 50% of the residents live below the federal poverty line. Watts also neighbors other poor communities. How will you communicate with the City of Los Angeles' emergency manager to coordinate an evacuation? How will you get the message out? Where will you take people? How long will they be able to survive without any help? The answers to these questions should be known long before a disaster strikes. Understanding these questions and answers is social intelligence, and then using the answers to build more competent plans is what we call real *jurisdictional competency*. In other words, having social intelligence for a specific region or area will provide you with critical jurisdictional competency.

JURISDICTIONAL COMPETENCY

Jurisdictional competency is achieved by developing socially intelligent emergency management systems. This means emergency managers, policy makers, and elected officials understand how the preexisting conditions in a community drive what happens during and after disaster. More importantly, once an emergency management system is socially intelligent and subsequently jurisdictionally competent, it understands how structural issues (social capital, poverty, government revenues, schools, demographics, etc.) are all connected. In understanding these connections, emergency managers can tailor emergency response and recovery resources in a manner that helps all people.

While uniform plans are important, knowing how to use your plans as a framework and then insert specific regional and local knowledge will be key in proficient planning, response, and recovery for a community. For example, jurisdictional competence will be very different from New Orleans to New York City to New Buffalo. All cities are on major bodies of water, all will rely on standard protocol and procedure, but an emergency manager in each community will need very different local information to prepare for, respond to, and recover from a disaster. Finally, jurisdictional competency is not rooted in living or working in a municipality—making intelligent and informed decisions during and after disaster involves asking the right questions. This means emergency managers must understand how social intelligence influences daily life in a community in order to ask the right questions during and after a disaster.

What you must also know is that social intelligence is not a tool to just understand socially vulnerable populations. We will detail how a disaster can often push individuals and families over the edge. That is, they were doing fine until the disaster struck—the impacts of the disaster on finances, support systems, and the family put them over the edge. There are thousands of communities that are vulnerable to events that fit this description. They are doing fine now, but if a disaster were to strike, these communities would fall over the edge. The economic downturn of 2008 has exposed many of these communities. A disaster event would be catastrophic for many of them.

While the nation will likely recover economically, we do not know the long-term impacts of this economic crisis. Many communities are providing fewer services due to the collapse of the real estate market and declining tax revenues. Poverty rates are climbing, and disasters are occurring more frequently. Emergency management must adapt to this reality because annual plan updates and template designs do not address the downturn. And unfortunately, adding more special plans (vulnerable populations, functional needs, etc.) will not work. What emergency management needs is a mechanism through which it can understand the real-time changes in a municipality to keep plans in a year-round living process. While it may be cliché, plans must truly become living documents that flex and adapt as necessary. The world is changing ever increasingly. If our processes and systems aren't cognizant of these changes, are inflexible, or can't otherwise accommodate them, we are doing the citizens we are charged to protect an injustice.

There are myriad problems that stress the ability of emergency management. Systemic problems like poverty severely stress planning, response, and recovery activities. But what often goes overlooked is the need for emergency management to understand social issues and problems. We argue that emergency management is perhaps the only profession that can objectively look at social issues and problems—we say this because the job of emergency management is not to fix these problems. That said, emergency management can play a role in helping government navigate the political landmines of social problems by removing information silos, coordinating information, and bringing resources to bear that aid in a more holistic understanding of the problem. But that has not happened thus far. There is still work that needs to be done, and we believe that our social intelligence model is a tool, one step toward a more robust emergency management field.

EMERGENCY MANAGEMENT APPROACH

We are often told that emergency managers are like the conductors of an orchestra. They can play some of the instruments but not all. What allows conductors to lead an orchestra is the ability to *understand* how all the instruments can work in unison in a musical movement. Similarly, an emergency manager understands how an emergency operations center (EOC) works. Emergency managers work with various entities within an EOC. They must understand how municipal services, state and federal agencies, nonprofits, private firms, and other responding entities operate and how they bring resources to bear. Emergency managers understand the connections between public services and their private sector partners. The entities that collaborate in an EOC are the emergency manager's orchestra. Emergency managers are the link between sanitation, public works, and the Red Cross. Simply put, they make systems work during times of crisis.

Emergency managers are coordinators of resources and information. If the field can understand a central coordinating authority is needed during and after a disaster, one would think emergency managers could coordinate discussions among EOC partners during the mitigation and preparedness phases.

Emergency plans are created and then left for annual updates. Communities ebb and flow in between plan updates and emergencies. City bond ratings are upgraded and downgraded. Budget deficits and surpluses come and go. Populations change. Companies open and close, and people change. All of these factors matter during and after an emergency. Why? Because the preexisting conditions in a community drive what happens

8

during and after a disaster. Yet emergency management plans and systems utilize a very narrow focus for community planning. That is, they don't necessarily factor the real-time ebbs and flows of a community into emergency planning. If we continue to wait for a disaster to strike to understand how communities change, there will continue to be a serious information gap. Information collecting and sharing is something that emergency management can do year-round. This is the essence of social intelligence.

MILITARY/PARAMILITARY STRUCTURES AND EMERGENCY MANAGEMENT

Emergency management is made up of firefighters, retired military, police, and other military or paramilitary organizations. These professionals have gone through strict and regimented training. In a military/ paramilitary organization, everyone has equal training and equipment, and then personnel are given specialized tasks and responsibilities. A military officer can assume than an enlisted soldier has had basic training. A firefighter or police officer can assume that his or her colleague has gone through the same, or similar, training regimen. Through that regimented training, a military or paramilitary unit achieves unit cohesion.

To a police officer, a criminal is a criminal. To a firefighter, a fire is a fire. A combatant in the theatre of war is the enemy to the military. First responders and military personnel have very specific duties as they relate to their job. In emergency management, responders must deal with a variety of populations. A hurricane or flood is not the same to every disaster victim, and emergency managers cannot treat these events like status levelers; their impact on populations varies by gender, race, socioeconomic status, etc. If the responding authorities are not aware of informal social networks, mistrust of government, or other key pieces of information, then the response and recovery efforts will continue to falter for lack of information.

Emergency management must evolve.

In addition to the law enforcement, fire, and military expertise and approach, emergency management must also involve sociologists, public health professionals, social workers, and other fields to help ask the right questions. Think about the wars in Iraq and Afghanistan. The coalition forces quickly realized that "shock and awe" and other shows of force only work for so long against enemy combatants. Instead, what works best is force and nation building. However, to engage in nation building, the forces had to understand the population's needs. That is, in addition to good training, our responders, policy makers, and emergency managers must understand daily life in the communities they serve. By attempting to understand the community, the military developed more informed strategies. These strategies were then leveraged via tactical units that informed operational units. This same approach could be utilized by emergency management and policy makers.

SITUATIONAL AWARENESS

According to the Federal Emergency Management Agency (FEMA) training unit response manual, the definition of situational awareness is:

> Situational awareness requires continuous monitoring of relevant sources of information regarding actual and developing incidents. For an effective national response, jurisdictions must continuously refine their ability to assess the situation as an incident unfolds and rapidly provide accurate and accessible information to decision makers. It is essential that all response partners develop a common operating picture and synchronize their response operations and resources.[*]

FEMA and emergency management understand that disasters are fluid and complex events. Situations can unfold throughout any given operation period. For example, during Hurricane Katrina, the nature of the response effort changed when reports of violence emerged from the Superdome. Rescue operations were ramped up when media outlets reported that thousands of residents were stranded on rooftops. As disaster events evolve, information is critical. Real-time information helps emergency managers, policy makers, and first responders make decisions. Information related to situational awareness can dictate policy decisions that would then inform tactical and operational decisions. This same thinking can be connected to our social intelligence framework, only on a more long-term, less crisis-motivated timescale.

The reality is that the landscape in any municipality can fluctuate and change daily. Imagine what a major factory closing can do to a community. Think about what impacts H1N1 could have had on vulnerable populations without access to primary care. Now think about how depressed tax revenue collection or a sagging economy can impact a fire

---

[*] Sourced from FEMA Training Manuals http://training.fema.gov/EMIWeb/Is/IS800B/SMs/04_IS800NRF_SM.pdf

or police department's ability to purchase equipment and make capital improvements. Emergency management's ability to help populations prepare, respond, and recover is directly tied to the fiscal health of the state and local government, the economic vitality of the municipality, and its municipal services.

A good example is the City of Chicago. In 2005, the City of Chicago took in over 10,000 evacuees from the Gulf Coast following Hurricanes Katrina and Rita. The resettlement was successful in that many evacuees were able to secure housing and jobs. If a similar disaster were to occur today, the City of Chicago might not be able to repeat its successes. Why? While there were lessons learned, the City of Chicago had a budget deficit of $522 million in 2010 and will have to address a $650 million deficit in 2011. Years of financial mismanagement have left the city strapped for cash. The real estate market has taken a significant hit and real estate transfer taxes (revenue stream tied directly to home sales and economy) are depressed, and with a rise in foreclosures, the city has had more difficulties in collecting property taxes. Private organizations and nonprofits that co-led the major initiative with the city are also experiencing challenges given the state of the economy. Private giving is down, and many smaller nonprofits are struggling to stay afloat. In short, the same robust response is likely not possible today.

To build on that last example, what we have also seen is the inability of emergency management to understand the full scope of long-term recovery. For example, after Hurricane Katrina housing, vouchers were issued to evacuees requiring affordable housing. The federal government informed evacuees that vouchers were portable. But what they failed to tell evacuees is that there are waiting lists for affordable housing in almost every major city. Waiting lists often stretch out into years. So regardless of whether the vouchers were portable, evacuees would have had no luck with finding housing.

Emergency management often relies too heavily on lessons learned or best practices. While a hot wash of every event can enable emergency managers and other responding authorities to catalogue successes and failures, a real-time assessment of situations is lacking. That is, situational awareness can vary from one operational period to the next—to understand those shifts or changes, the National Response Framework is built to be scalable and flexible. However, prior to disaster, in many cases emergency management has not harnessed the full potential of the National Response Framework and Incident Command System to understand how communities change in real time.

11

SITUATIONAL AWARENESS—RESPONSE AND RECOVERY

After Hurricane Katrina, there was a movement within emergency management to reinforce the statement "all response is local." That statement also led federal, state, and local authorities to start educating the public on emergency preparedness in the home and communicating that individuals have a responsibility to sustain themselves for 72 hours after a disaster. But imagine how difficult it is for a single mother, or an elderly couple on a fixed income, or a middle-class family that has a spouse that has been unemployed, or a struggling student to purchase a go-kit. Emergency go-kits cost approximately $100. For many Americans, investing in these kits for an unlikely emergency when dealing with daily crises is an unrealistic expectation. Going beyond the kit, emergency management must also understand that go-kits need to be reinforced with training and mitigation procedures. All of these processes are time-consuming and costly. Access to resources is critical in the preparedness phase of emergency management. Absent these resources, what is emergency management's plan to help prepare socially vulnerable populations?

The answer lies in our ability to collect information, analyze it, and then build robust emergency management systems that address this reality. This will require the field to develop situational awareness in real time. Why? Because preparing for disaster is the furthest thing from a person's mind when dealing with the day-to-day disasters of survival.

We have already run through a few examples, but consider this one:

> Imagine if you were told to evacuate your city within the next 24 hours. What would you do? If you are reading this book, we will assume that you have access to credit, cash, and support systems (family and friends). We are making these assumptions without knowledge of your circumstances. These are some very big assumptions—emergency management makes these assumptions all the time. So you have access to resources that are critical during an evacuation and recovery. Let's go over these resources:
>
> > Credit: If you had to book a flight, rental, car, or hotel you would need a credit card.
> > Cash: If you have access to credit, then chances are good that you have access to some cash.
> > Support systems: If you know that you can leave and stay with family or friends, and have access to cash/credit, you will likely heed evacuation orders and come back when it is safe.

When that evacuation order is given, you will likely hop in your car or rent one. Then you might get a last-minute flight or book a hotel in a safe area. You would then rely on support systems and your available cash to carry you through the 72 hours. Now if you had all of these resources, you would also probably be able to purchase a go-kit in advance of the disaster. If you had access to all of the resources named above, you likely live in an area where you have had access to good schools, banks, and grocery stores. It is also likely that you have a safety net that can carry you over until any FEMA or disaster aid arrives.

Is it dangerous to make all of these assumptions? It is dangerous because, again, we don't know your circumstances, but our emergency plans assume that you are resilient in this way. Now consider this scenario:

Imagine if you are married and you have two young children living in a small town in Ohio. You and your spouse both work in the same meat packing plant. You both relocated to the area several years ago, and while you have friends, you do not have immediate family in the area as a support structure. An economic downturn forces your company to lay off 30% of its workforce. You are spared but your spouse is not. Your costs for health insurance spike and you drain your $5,000 savings to pay the monthly COBRA. Because of the mass layoffs, stores in the community start to close and prices for normal goods increase sharply. Between your income and your spouse's unemployment check your family barely gets by. After 4 months, you start to worry about your job. You start to look in neighboring towns for work. Because of the layoffs affecting your neighbors, more and more people default on their mortgage payments. Your home value plummets because of the drop in value across town. The municipal government struggles to provide services because of declining revenues. As a result, you are upside down on your mortgage and after 6 months, your spouse's unemployment benefits run out. What will you do?

You may be wondering what this last example has to do with emergency management. For a long time, emergency management has asked us the same thing. What has happened in this scenario is occurring more and more nationwide. Think about what would happen in this town if it were located in tornado alley and a EF*4 tornado were to strike. Or,

---

* Enhanced Fujita scale.

13

imagine if this community were on the Mississippi River during the flooding in 2008. A disaster would exacerbate an already deteriorating situation, and, in most cases, push a family from eking out daily survival to unable to get by. Resources would be entirely expended.

Hurricanes Katrina and Rita simply exposed these systemic social problems. The reality is that not many communities are made up fully of people from the first example—those that can access additional resources in an emergency. That is, unless you are talking about communities like Malibu, California. American towns and cities are increasingly diverse. We are a melting pot of races, cultures, incomes, political beliefs, etc. So emergency management cannot make broad assumptions—we do credit FEMA and local emergency management agencies with recognizing a need to plan for functional/accessible needs populations, pets, vulnerable populations, and children. But far too often, these special planning groups are planned for without fully understanding their circumstances prior to disaster. Furthermore, recovery for these groups varies drastically from those that are able to recover on their own.

So what can emergency management really do? There first needs to be recognition that the field must be more aware of what is happening in the community. That, at a minimum, is a start. Emergency managers must understand, in at least a general, basic sense, how a factory closing can impact tax revenue. They must understand how that impacted tax revenue impacts the fire department's ability to purchase equipment and hire grant writers or firefighters. Emergency managers must understand how that factory closing can impact the community and its support services. They must have an understanding of what social service systems can be brought to bear. This is real-time situational awareness.

The vast majority of your time as an emergency manager will be spent in planning, preparedness, and mitigation. While major disasters are occurring more frequently and their impacts are more severe, most emergency managers will likely deal with smaller events. Most of our careers are spent preparing for the big disaster. Emergency management should be spending that time wisely. Waiting for disaster to strike before understanding the interconnections across departments, social structures, and communities is irresponsible. The ability to understand your community is already there. Will you harness available information to become socially intelligent?

2

Disaster Case Studies

How can we learn from mistakes made during emergency response and recovery? Look no further than to history. This book frequently refers to the $1 trillion spent on disaster response and recovery. And this book repeatedly states that communities are no better prepared for the next disaster after having spent these dollars. This chapter will focus on various disasters spanning the 20th and 21st centuries. Each case study will provide background on the disaster event, impacts, and lessons learned.

The case studies will examine the following disasters:

- San Francisco Earthquake
- Hurricane Andrew
- Northridge Earthquake
- Hurricane Katrina
- Hurricane Katrina evacuation to Chicago
- Great Flood of 1927—Mississippi River
- Chicago heat wave of 1995
- Illinois floods of 2010
- *Deepwater Horizon* oil spill of 2010
- Hurricane Sandy

The 10 case studies reflect consistent issues. The impacts of each disaster were exacerbated by preexisting conditions. And during each disaster, preexisting conditions were largely ignored for a template emergency response. Some of the preexisting conditions include

- Poverty
- Lack of affordable housing

- Discriminatory behavior toward minority communities
- Lack of access to government services
- Lack of trust in government

Beyond the consistent themes, this chapter will help frame the social intelligence theory. That is, it will help emergency managers begin to clearly link historical context with preexisting conditions. Emergency management has only recently begun to pay close attention to and understand the connection between socioeconomic systems and disaster. The tendency has been to divorce everyday life from the challenges faced during and after a disaster. We focus on every person's equal exposure to a storm or event as we know that something like a tornado or oil spill will not discriminate in its path of destruction. We will all face Mother Nature equally—we will all face major losses.

Some, however, will be better equipped for response and recovery. They will have insurance, disposable income or savings, access to credit or strong social networks, etc. This will not make them immune to the effects of an event, but will likely stave off the devastation that hits those with less stability and fewer resources, and therefore much less ability for resiliency. Again, disasters are not status levelers. This, of course, is not just limited to individuals. Resiliency of a community or neighborhood also cannot be divorced from its everyday reality. Many factors contribute to the successes and challenges a community faces in preparing for, responding to, and recovering from a disaster. The challenges are much greater depending on preexisting conditions. For example, a flood that hits the Gold Coast neighborhood in the City of Chicago will likely look much different than the same event less than 3 miles away in the Lawndale neighborhood. One neighborhood is wealthy; the other is not. Both neighborhoods will have people with functional and accessible needs, but those from a poorer community will likely have less ability to recover in the long term given preexisting conditions that relate to access to resources and ultimately resiliency.

This separation of everyday struggles from those faced post-disaster is most certainly related to emergency management's focus on addressing the needs caused solely by a disaster. For example, it is not the duty of Federal Emergency Management Agency (FEMA) to make a person or family whole after an event. Aid is available to compensate for certain needs as a result of an event, but not all needs are eligible for assistance. In attempting to separate what falls under the purview of emergency management and what does not, we may have unnecessarily separated the field from critical situational awareness.

We know that the United States has faced disasters caused by humans and Mother Nature since its inception. We also know that we are seeing more events with more catastrophic results and on a much more regular basis. And, we continue to regularly spend hundreds of millions, and in many cases billions, of dollars to respond to and recover from disasters. We estimate over $1 trillion spent alone in the last three decades.

We will most certainly continue to face disasters on a regular basis. We will also continue to be challenged by many of the same issues in each and every disaster, especially as it relates to preexisting conditions of a family or community. And, again, while emergency management cannot eradicate problems like poverty or other socioeconomic inequalities, it can be intelligent about the community. And not just in understanding that vulnerable populations (or functional and accessible needs populations) exist. While knowing that is an important start, being truly socially intelligent, or having what we term jurisdictional competence about one's community, means knowing that poverty exists, at what rates, and what that means in one's specific community—or, what that truly looks like from day to day. Rather than a sole focus on resources, or a sole focus on the disaster event itself, social intelligence is what will truly help inform us on decisions that can be most impactful for a given community.

For example, as we see disaster planning better incorporate lessons learned from recent disasters, we see something like the presence of poverty in communities get attention. In order to figure out what this means as it relates to disaster, different types of assessments are taking place nationwide to capture important information. Specifically, disaster planners now know that many people in Katrina did not evacuate, not because they did not heed warnings, but because they could not due to lack of access to transportation. So, as a result, we see plans gathering data on car ownership, the assumption being that if cities can know how many people have cars, they can know how many people can self-evacuate. While this is an important step in better understanding the capabilities of a community with high rates of poverty, it is problematic because for many, owning a car does not equal the ability to self-evacuate. Because many people will move in and out of poverty over their lifetime, the ability to use one's car, say by having enough money to fill the gas tank regularly, will change according to the current financial situation. So, as in the case of New Orleans, we know that many people had cars; they just did not have disposable income at that time in the month, could not fill a gas tank, and therefore could not self-evacuate. In this case, knowing what poverty looks like in a community, its prevalence, and its rates will be critical in determining how to best plan, respond, and recover. And, as we look more toward personal preparedness and making sure families can prepare themselves, we must consider the interplay between being low-income and the ability to prepare, respond, and recover.

What we can see from an overview of just 10 events in the last 100 years is that the same types of issues will be present over and over again—in all phases of disaster. However, how we prepare for, respond to, and recover from disasters can be much more comprehensive and less costly if we have better awareness of how something like poverty or limited language ability or disability or housing markets or municipal budgetary concerns interact with all systems in a community. Issues will continue to be the same, but communities are unique and constantly changing—emergency management must have the kind of situational awareness that allows for even greater flexibility in response and recovery capabilities.

DISASTER EVENT: SAN FRANCISCO EARTHQUAKE (1906)

What Happened?

On April 18, 1906, at 5:12 a.m., a foreshock with significant force was felt throughout the San Francisco Bay Area—the earthquake then broke loose approximately 20–25 seconds later. The epicenter was near San Francisco, and violent shocks were felt intermittently throughout the strong shaking, which lasted about 45–60 seconds. The earthquake was felt from Oregon in the north, to south of Los Angeles, and inland as far as the center of Nevada. Rupturing occurred along 296 miles of the San Andreas Fault.* As tremors subsided, fires broke out as a result of downed power lines, and winds swept flames over the city. Fire hoses were rendered useless as water mains were destroyed from the earthquake, which allowed fire to rage for 3 days. Over 4 square miles of the city were burned before the fire burnt itself out. People fled to the suburbs, countryside, and into Golden

---

* Survey, U. (2011). San Francisco Earthquake of 1906. Retrieved from http://www.eoearth. org/view/article/164914

Gate Park. Military troops were called in to maintain order and provide initial relief.[*]

Impact

As a result of the disaster, over 3,000 people were killed and approximately 225,000 people were left homeless. The population of San Francisco at the time was about 400,000, leaving over half of the city's residents without a place to live. The uncontrolled burning that followed the earthquake seemed to cause more damage than the earthquake itself, leaving 4.7 square miles burned. Over 28,000 buildings were destroyed (24,671 wood, 3,168 brick), and the estimated property damage totaled about $400 million in 1906 dollars—approximately $9.8 billion adjusted for 2010.[†]

Challenges to Recovery

Following the devastating earthquake, and over a 3-day period, nearly three-quarters of San Francisco burned to the ground. Over 50% of San Franciscans were left homeless. Recent estimates show that nearly 90% of the city's building structures were destroyed.[‡] Like many of the case studies discussed in this section, the disaster event was near catastrophic, but not all people were impacted equally.

One of the first challenges was to restart government services. Widespread looting and chaos reigned in the immediate aftermath of the earthquake. Compounding this, fires roared for many days and destroyed nearly every building in San Francisco. To regain control of the city, the military stepped in and patrolled the city for over two months. The earthquake destroyed city hall, and with it nearly every municipal record. Records included birth certificates, property titles and deeds, and documents related to city finances. Built with shoddy materials, including old newspapers and trash, the building quickly crumbled. The original

---

[*] Henderson, Andrea, The Human Geography of Catastrophe: Family Bonds, Community Ties, and Disaster Relief After the 1906 San Francisco Earthquake and Fire. *Southern California Quarterly*, 88(1): pp. 37–70, Spring 2006. Published by: University of California Press on behalf of the Historical Society of Southern California Article DOI: 10.2307/41172296. Article Stable URL: http://www.jstor.org/stable/41172296. Military issues first mentioned on page: 50—the military efforts are mentioned many times throughout the article

[†] Survey, U. (2011). San Francisco Earthquake of 1906. Retrieved from http://www.eoearth.org/view/article/164914

[‡] Potter, Sean, April 18, 1906: The Great San Francisco Earthquake, *Weatherwise* 61(2): 14–15, 2008. Academic Search Complete, EBSCOhost (accessed August 27, 2013).

construction of city hall was mired in corruption, and much of the funding toward the construction of the building was pocketed by contractors. The destruction of city hall and the discovery of shoddy materials and workmanship were the symbolic underbelly of San Francisco. In fact, in December 1906, the mayor of San Francisco and a leading labor leader would be indicted for corruption.* President Roosevelt, through proclamation, asked that financial aid be channeled through the Red Cross in order to bypass the city government, whose honesty and management ability he questioned.† With city hall gone, San Francisco officials quickly set up a government command post, but local government on its own was not able to serve its people. The result: with the Army providing general security throughout the city, operational and logistical support for food supply chains, and creating "refugee" camps for shelter, government was able to partner with a committee of business and civic leaders to jump-start recovery.

> The municipal government was displaced the day of the earthquake by a citizens committee of business and civic leaders. This Committee would control local government funds, including $10 million in donations, and dictate or cajole liberal land use, zoning, business licensing, and building trade rules to speed redevelopment and build confidence in the recovery.‡

To summarize, these were the challenges facing people and officials in San Francisco:

1. Most of San Francisco was destroyed during the earthquake or in the fires afterward.
2. There was no running freshwater or sewage services.
3. Food supplies and storage facilities were destroyed.
4. Nearly every government building and record was destroyed. This made recovery for Asian populations especially difficult. Prior to the earthquake, Asians not born in the United States were unable to own property. With city hall destroyed, all birth certificates were destroyed.

---

* Coate, Douglas, *Disaster and Recovery: The Public and Private Sectors in the Aftermath of the 1906 Earthquake in San Francisco*, Rutgers University, Newark, NJ, July 2010.
† Ibid.
‡ Ibid.

5. The public safety departments in San Francisco were over-whelmed or incapacitated. To provide security in the city, the U.S. Army deployed to it.

6. Critical infrastructure was destroyed by the earthquake or result-ing fires. In the immediate aftermath, there was no potable water available for the remaining population and for those involved in recovery efforts. To address this, the Army pumped in seawater to put out fires, brought in drinking water, and built refugee camps for sheltering the homeless.

7. A committee of fifty (leaders in the public and private sector) came together to marshal resources and spearhead recovery in San Francisco.

8. Recovery funds and grants were disbursed very slowly. The ardu-ous process made it difficult for vulnerable populations to start the recovery process.

9. Looking a bit deeper, the City of San Francisco had deep divides among its population. While a significant number of Chinese and Chinese Americans and Japanese and Japanese Americans lived in San Francisco and built thriving communities, city gov-ernment did not do much to create a culture of acceptance. "We favor the absolute exclusion of all Asiatics—Japanese as well as Chinese," said the platform of the Workingman's Party, which controlled city government.[*] Women were not able to vote, and only Caucasians were allowed to own property.[†]

Lessons Learned

Much was learned regarding earthquake science as a result of this event. And, based on earthquake modeling, the best guess is that a 1906-type earthquake occurs at about 200-year intervals. According to the U.S. Geological Survey (USGS), due to the time needed to accumulate the kind of movement needed to equal a 20-foot offset, there is only a 2% chance that such an earthquake could occur in the next few decades. The real threat to San Francisco Bay is from somewhat smaller earthquakes (about a magnitude of 7 on the Richter scale) coming from the Hayward, Rodgers Creek, or San Andreas Fault. A quake of this size is considered likely to

---

[*] Nolte, Carl, Nothing Magical about City before 1906 Quake/Town Was Smoggy, Corrupt and Dirty, *San Francisco Chronicle*, Sunday, April 18, 2004 (accessed online August 27, 2013).
[†] Ibid.

occur before 2032.[*] Consider also that San Francisco's and the surrounding suburbs' current population is roughly 825,000.[†]

This ever-present geological threat, coupled with the also immediate threat of state insolvency, places California in a very tenuous situation. While we know California to be among the worldwide experts in disaster preparedness, especially as it relates to earthquakes, the current budget crisis diminishes preparedness, response, and recovery capabilities. Individual resiliency is already weak in vulnerable communities, and the state of the economy has even more people living in tenuous everyday situations. In addition, communities are making serious cuts to critical programming and infrastructure, which have profound impacts on everyday life for Californians, even more so as it relates to disasters, as the preexisting conditions in communities drive what happens during and after disasters.

Preexisting social and social issues are a common theme in this book, and these issues were a constant in preearthquake San Francisco. The earthquake brought many of these preexisting issues to the surface. Asian populations suffered great discrimination preearthquake. While the earthquake exposed racist policies that already existed and predated the disaster, these policies were exacerbated by the earthquake. With nearly every document destroyed in the city hall collapse, Asian Americans had a difficult time accessing services and starting the recovery process. Without documentation to prove they were born in the United States, property ownership was impossible. Women were unable to vote and exercise any real authority during the recovery. On top of all this, a corrupt local government complicated disaster relief efforts. Are there any common themes between the earthquake and Katrina?

1. Lack of trust in government due to corruption.
2. History of racist and discriminatory policies exacerbates recovery efforts for minority populations.
3. An unwillingness to connect preexisting conditions in a community with what happens after disaster.

While the San Francisco Earthquake took place over 100 years ago, we can still see many of the same recovery issues today. An attention to getting people safely housed and rapidly reemployed is critical to resiliency of a family and a community. Also important is ensuring that recovery

---

[*] Survey, U. (2011). San Francisco Earthquake of 1906. Retrieved from http://www.eoearth.org/view/article/164914

[†] http://quickfacts.census.gov/qfd/states/06/06075.html

funds are quickly dispersed, and that paying close attention to embedded social issues is key. In San Francisco, decades of exclusionary and discriminatory policies toward Asians complicated recovery efforts. While San Francisco and the United States have come a long way in the last century, the remnants of these policies remain today in various forms. Finally, knowing that timely access to disaster recovery funds often determines the overall success of recovery efforts is crucial. After the 1906 earthquake, accessing disaster recovery funds required claimants to navigate a slow and arduous process. Many of the issues outlined in "Challenges to Recovery" would be a factor today.

DISASTER EVENT: HURRICANE ANDREW (1992)

What Happened?

Hurricane Andrew began as a storm off the west coast of Africa on August 14, 1992. By August 16, the weather system was upgraded to a tropical depression, and by the next day, Tropical Storm Andrew had formed. The tropical storm strengthened, and by the morning of August 22, Hurricane Andrew formed. By late on August 23, the hurricane reached category 4 status as it passed over the eastern Bahamas. Andrew then crossed the Florida Straits into south Florida near Homestead. Major damage was caused to south Florida, especially south of Miami. After only a few hours over south Florida, the hurricane turned north and made a second landfall in Louisiana. At its peak, Hurricane Andrew packed maximum sustained winds of 165 mph, with peak gusts of more than 200 mph, making it only one of three such storms to hit the U.S. coastline.[*] As Andrew bore down on south Florida, officials ordered large-scale evacuations. About 55,000 people left the Florida Keys, and an additional 517,000 people in Dade County were told to leave.[†]

Impact

The category 5 storm caused 43 deaths and approximately $30 billion in damages (1992 dollars), making it the costliest disaster in U.S. history at the time. Unusually, wind, not storm surge or rain, was the most destructive

---

[*] http://www.tropicalweather.net/hurricane_andrew_facts.htm
[†] http://www.sun-sentinel.com/news/weather/hurricane/sfl-scane22aug22,0,1661757.story

Figure 2.1   Dade County, Florida, August 24, 1992. An aerial view showing damage from one of the most destructive hurricanes in America's history. One million people were evacuated and 54 died. (From FEMA.)

force. Only about 7 inches of rain fell on Florida. When it was all over, more than 250,000 people were left homeless, 82,000 businesses were destroyed, 86,000 people lost their jobs, and approximately 100,000 residents of south Dade County permanently left. Violent winds destroyed more than 25,000 houses and damaged more than 100,000 others (Figure 2.1). Thousands of cars were totaled, 15,000 boats (Figure 2.2) were destroyed, and more than 150 airplanes were also destroyed. In the Everglades, 70,000 acres of native mangrove trees were decimated, and 33% of the coral reefs at Biscayne National Park, as well as 90% of South Dade's native pinelands, were severely impacted.

In addition, the entire response was botched by all levels of government. Initial forecasts did accurately predict the severity of the storm. In fact, most forecasts predicted a severe tropical storm or a hurricane that would break apart once it reached shore.* However, once it reached shore as a category 5 hurricane, many monitoring stations and instruments

---

* Hughes, Tracy, The Evolution of Federal Emergency Response since Hurricane Andrew, *Fire Engineering* 165(2): 90–94, 2012. Academic Search Complete, EBSCOhost (accessed September 13, 2013).

Figure 2.2 Dade County, Florida, August 24, 1992. An aerial view showing the damage to a local marina. (From FEMA.)

were destroyed by the force of the storm. With the storm came a 14-foot surge, heavy rains (topping over 7 inches), winds, and flooding.[*] Because the storm missed much of Miami, many people believed the damage was not severe. As a result, many county and state officials were slow to respond. In addition, and after the National Guard was deployed to maintain security and prevent looting, the governor failed to provide a written request (as required) seeking federal assistance. When the initial impacts of the storm were combined with the failure of all levels of government to respond in a timely manner, the recovery was delayed unnecessarily.[†,‡]

Challenges to Recovery

Nothing sums up the challenges to starting the recovery process like the following quote:

[*] Hughes, Tracy, The Evolution of Federal Emergency Response since Hurricane Andrew, *Fire Engineering* 165(2): 90–94, 2012. Academic Search Complete, EBSCOhost (accessed September 13, 2013).

[†] http://www.nationalgeographic.com/forcesofnature/forces/h_3.html

[‡] http://www.sptimes.com/2002/webspecials02/andrew/

Kathleen "Kate" Hale, the emergency management coordinator for Dade County, called a press conference and, with tears in her eyes, asked, "Where in the hell is the cavalry on this one? They keep saying we're going to get supplies. For God's sake, where are they?"[*]

Because the storm missed much of the Miami metropolitan area, the assumption among many officials was the storm had not been as damaging as previously thought. In fact, most believed the storm landed as a tropical storm. Unbeknownst to them, the storm landed as a category 5 hurricane, and with it brought a wave of catastrophic proportions.

These days have been engrained in the minds of all South Floridians who rode through the storm as residents spent days sweating, crying, and questioning how to move on. Looting soon became a major problem and residents had to guard their property, few remaining possessions, and, most importantly, safe food and water, with guns. Help was slow to arrive, partly a result of poor planning, partly a consequence of insufficient communication, and at least to many residents of South Dade County, largely because the government did not care about them.[†]

The day after Hurricane Andrew made landfall, news crews and people in the Miami metro area rejoiced—the hurricane spared their communities. As the day wore on, images and news of the devastation south of Miami started to filter in. This type of destruction had not been seen before. Entire neighborhoods were unrecognizable, even to residents. Homes had been ripped to shreds and there were virtually no signs of life.[‡] And as people emerged from the devastated communities, they quickly realized that there were no provisions for electricity, water, or food. Because the storm leveled so many homes, shelter from the broiling sun was another immediate concern. The hopelessness of the situation sparked widespread looting. To restore security, the National Guard was called in to control looting, and eventually every branch of the U.S. armed forces was deployed to help clean up and restore order. After security was restored, the economic and social impact to Homestead and other impacted communities was near catastrophic. Overnight, Homestead Air Force Base, which employed thousands of area residents, was destroyed.

---

[*] Mathews, T., and P. Katel, What Went Wrong? *Newsweek* 120(10): 22, 1992. Retrieved from Academic Search Premier database.

[†] Kessner, Jeffrey L., Racial and Ethnic Conflict in South Florida: Hurricane Andrew and the Housing Crisis by Class of 2007, Wesleyan University. http://wesscholar.wesleyan.edu/cgi/viewcontent.cgi?article=1001&context=etd_hon_theses

[‡] Ibid.

Like other communities studied in this chapter, the disaster event created a cascade of economic and social impacts.

Included among the challenges to recovery:

- Anger and outrage after long delays to reestablish security in the disaster-impacted area led to mistrust and anger toward local government authorities and other organizations.
- Because security efforts were delayed, impacted people believed they were purposefully excluded from response efforts, leading to lack of trust in government.
- As a result of the hurricane, the City of Homestead suffered a 31% decline to its population, 60% of the aggregated residential property value, and 29% of its average commercial real estate value.[*]
- With Homestead Air Force Base decimated, local employment at the base shrunk. As a result, the tax base in Homestead and the affected region was impacted. Many industries (e.g., hospitality, retail, and tourism) were impacted by the hurricane. The loss of the Air Force base compounded the initial losses.
- The hurricane rendered 160,000 people homeless.[†]
- As minority communities were disbursed around the area and region, family and neighborhood networks were disrupted.
- Tent cities were established by the American Red Cross. In a study of the tent cities that were established as temporary housing following Hurricane Andrew, (Yelvington 1997) reported that a number of post-traumatic stress disorder cases were evident. Many camp residents, most of whom were poor or working-class ethnic minorities, reported psychological depression and strained familial relations.[‡]
- Many families were placed in FEMA trailers next to their homes. While thousands qualified for FEMA trailers, many families were denied. Why? FEMA had strict guidelines that favored small nuclear families with a single head of household.[§]
- Lower-income individuals and families were in need of resources long after FEMA deadlines had expired. Conversely, individuals

---

[*] http://www.gao.gov/new.items/d09811.pdf.

[†] http://archive.gao.gov/t2pbat5/149631.pdf.

[‡] Fothergill, Alice, and Lori A. Peek, *Poverty and Disasters in the United States: A Review of Recent Sociological Findings.*

[§] http://www.academia.edu/4313624/Stretching_the_Bonds_The_Families_of_Andrew, p. 154.

and families with higher incomes had the knowledge to navigate disaster relief systems to access aid.[*]
- Owner-occupied units were faster to recover than rental units.[†]
- Minority neighborhoods were much slower to recover during the rebuilding process than mostly white neighborhoods.[‡]
- Minority households' limited economic power and political representation make them less likely to have input into planning for disaster recovery programs and activities.
- Activities.[§] Wind damage exposed shoddy construction across the impacted area.

It is clear the same issues arise during most disasters. Vulnerable populations (minorities, low-income, etc.) are most adversely affected. Failing to understand the pre-existing conditions will guarantee the same failures during and after each disaster.

Lessons Learned

Like other disasters studied in this book (and most others not covered in this book), the same issues pop up during and after each disaster. The response for Hurricane Andrew was botched from the start. Because of the botched response, federal officials began to develop systems for proper command and control during and after disaster scenarios. The National Incident Management System (NIMS), the National Response Plan (NRP), and the current National Response Framework (NRF) all are rooted in the failed Hurricane Andrew response. While disaster policies evolved, and were aimed at addressing, the failed Andrew response, the NRP failed after Hurricanes Katrina and Rita. The failure of the NRP led to the creation of the NRF. Emergency management and disaster-related policy continues to evolve, but these "paper" plans do little to acknowledge the on-the-ground realities in communities. So while it is safe to say that emergency management is learning some from each failed response, the larger policy approach has not changed. This is where we are today.

---

[*] Seidenberg, Jennifer, Cultural Competency in Disaster Recovery: Lessons Learned from the Hurricane Katrina Experience for Better Serving Marginalized Communities.

[†] Zhang, Yang, and Walter Gillis Peacock, Planning for Housing Recovery? Lessons Learned from Hurricane Andrew, *Journal of the American Planning Association* 76(1): 5–24, 2010. Academic Search Complete, EBSCOhost (accessed September 3, 2013).

[‡] Ibid.

[§] Ibid.

While some of the more global response mechanisms in emergency management evolved (and later failed during Hurricane Katrina), there were concrete improvements and lessons learned from this hurricane. Some include

- Building codes were strengthened to ensure structures can withstand hurricane force winds; the State of Florida developed a state building code in 2002.
- Better technology was developed surrounding hurricane detection and prediction.
- Every Floridian is now encouraged to develop a personalized hurricane plan and maintain basic survival supplies for up to 1 week.

Perhaps the most important lesson learned from Hurricane Andrew is that long-term recovery is a reality government should contemplate. Thirteen years after Hurricane Andrew, Katrina struck the Gulf Coast. And nearly 8 years after Katrina, some impacted communities in the Gulf Coast are still recovering.

DISASTER EVENT: NORTHRIDGE EARTHQUAKE (1994)

What Happened?

On Monday, January 17, 1994, an earthquake with the magnitude of 6.7 on the Richter scale hit the Los Angeles area. The earthquake lasted roughly 15 to 20 seconds, relatively short in duration, yet caused one of the most costly disasters in U.S. history.[*] Its epicenter was located beneath the San Fernando Valley, in Northridge, which is about 17 miles west-northwest from downtown Los Angeles.[†] While it was fortunate that the disaster occurred in the early morning hours of a holiday weekend (4:31 a.m. on the observed MLK Jr. Day) when highways had few motorists and other vulnerable structures were vacant, the earthquake still inflicted major tolls.[‡]

Impact

Over 60 people were killed, about 1,500 were admitted to hospitals with major injuries, and an additional 16,000 were admitted and released. The

---

[*] Bolin and Stanford (1998) *The Northridge Earthquake*. Routledge, London.
[†] http://resilience.abag.ca.gov/wp-content/documents/resilience/toolkit/The%20 Northridge%20Earthquake%20and%20its%20Economic%20and%20Social%20Impacts.pdf
[‡] Bolin and Stanford (1998) *The Northridge Earthquake*. Routledge, London.

Figure 2.3 Northridge Earthquake, California, January 17, 1994. Buildings, cars, and personal property were all destroyed when the earthquake struck. Approximately 114,000 residential and commercial structures were damaged, estimated at $25 billion. (From FEMA.)

Los Angeles metropolitan area had a population of approximately 10 million inhabitants when the Northridge Earthquake hit. Direct economic losses from the earthquake are estimated at over $25 billion.[*] An additional $12.5 billion was spent by private insurance companies to compensate for the damage, totaling upwards of $37.5 billion in disaster-related costs[†] (Figure 2.3).

The Northridge Earthquake produced the strongest ground motions ever recorded in a North American urban setting, causing structural damage as far away as 52 miles.[‡] Estimates of the number of people temporarily or permanently displaced due to home damage ranged between

---

[*] Mileti, Dennis. 1999. *Disasters by Design:: A Reassessment of Natural Hazards in the United States.* Joseph Henry Press. Washington D.C.

[†] http://resilience.abag.ca.gov/wp-content/documents/resilience/toolkit/The%20 Northridge%20Earthquake%20and%20its%20Economic%20and%20Social%20Impacts.pdf

[‡] Carino, Nicolas J., Chung, RM, Lew, Hai S., Taylor, A W, and Walton, William D. Northridge Earthquake: Performance of Structures, Lifelines and Fire Protection Systems (NIST SP 862). May 1, 1994. Available at: www.nist.gov/manuscript-publication-search. cfm?pub_id=908750. Published by the US Department of Commerce. pp. 1–2.

80,000 and 125,000.[*] As of the end of October 1994, the Federal Emergency Management Agency had received 519,000 applications for disaster housing assistance. In the 3 years that followed, requests for assistance topped more than 681,000 applications.[†] The earthquake also did extensive damage to public buildings, roads, bridges, and water control facilities.[‡] About 11 major roads and freeway interchanges as well as 12,500 housing units totally collapsed or suffered severe damage.[§]

Challenges to Recovery

In California and especially around Los Angeles, people of all backgrounds are exposed to risks associated with wildfires, earthquakes, and floods. General exposure to such hazards cuts across class, ethnic, and racial lines in the region.[¶] That is, in many disaster-impacted areas, socially and economically vulnerable populations are concentrated in areas where environmental vulnerabilities are most severe. In Los Angeles, vulnerability is more irregularly distributed than the norm.[**] The earthquake destroyed over 49,000 housing units, and the Red Cross was housing up to 7,000 per night at the height of the crisis.[††]

And as a result, in the aftermath of the Northridge Earthquake, housing and housing-related issues became significant challenges—especially as related to affordable housing and those who needed it. The availability of affordable housing for those on low incomes, particularly Latinos, the elderly, and farm workers, was an immediate concern. At that time

---

[*] Carino, Nicolas J., Chung, RM, Lew, Hai S., Taylor, A W, and Walton, William D. Northridge Earthquake: Performance of Structures, Lifelines and Fire Protection Systems (NIST SP 862). May 1, 1994. Available at: www.nist.gov/manuscript-publication-search.cfm?pub_id=908750. Published by the US Department of Commerce. p. 1.

[†] Carino, Nicolas J., Chung, RM, Lew, Hai S., Taylor, A W, and Walton, William D. Northridge Earthquake: Performance of Structures, Lifelines and Fire Protection Systems (NIST SP 862). May 1, 1994. Available at: www.nist.gov/manuscript-publication-search.cfm?pub_id=908750. Published by the US Department of Commerce. p. 2.

[‡] http://udspace.udel.edu/handle/19716/633

[§] Carino, Nicolas J., Chung, RM, Lew, Hai S., Taylor, A W, and Walton, William D. Northridge Earthquake: Performance of Structures, Lifelines and Fire Protection Systems (NIST SP 862). May 1, 1994. Available at: www.nist.gov/manuscript-publication-search.cfm?pub_id=908750. Published by the US Department of Commerce. p. 1.

[¶] Bolin, Robert, and Lois Stanford, The Northridge Earthquake: Community-Based Approaches to Unmet Recovery Needs, *Disasters* 22(1): 21, 1998. Academic Search Complete, EBSCOhost (accessed September 14, 2013).

[**] Ibid.

[††] Ibid.

31

Figure 2.4 Northridge Earthquake, California, January 17, 1994. Residents return to retrieve belongings from a damaged building. (From FEMA.)

many communities with close proximity to Los Angeles had seen a steady upward pressure on house prices and property values, making affordable housing increasingly scarce. In addition, the earthquake strained the housing market even further as hundreds of lower-cost housing units were damaged or destroyed, and the more expensive surplus of housing was taken by those who could afford it* (Figure 2.4). With over 519,000 people seeking disaster assistance, access to affordable housing was going to be the linchpin in a successful recovery effort. In the area surrounding the earthquake-impacted zone, affordable housing was scarce. With so many units destroyed and many more people seeking affordable housing than units available predisaster, a housing crisis was sure to emerge.

Specific challenges for housing issues post-disaster included

- Many working-class communities also sustained commercial damage in the earthquake, which resulted in no means of income for many for prolonged periods of time.
- California is home to a significant migrant worker population, many of whom are housed in labor camps and near other

---

* Bolin and Stanford (1998) *The Northridge Earthquake*. Routledge, London.

structures of production. This is a population that is largely invisible in official population and poverty figures. Of the migrant workers, there is also an unknown number of workers in the country without documentation.[*]

- Many recent immigrants, mostly Mexican immigrants, were lacking experience with disasters and a federal aid system and were therefore not familiar with how to access the system. In the chaos and devastation, it became difficult to obtain information, and many were therefore reluctant to seek out public resources due to unfamiliarity with and some fear of government systems. This fear was partially fed by an anti-immigrant political scene that ultimately kept people from accessing needed aid. Anti-immigrant discourse was a prominent feature of the political scene at the time.
- While the Red Cross did have a continuing assistance program to help people after their FEMA grants had been exhausted, or if they were ineligible, due to funding shortfalls, longer-term unmet needs had to fall to other nongovernmental organizations (NGOs) or go unaddressed.[†]
- Resistance to rebuilding low-income housing. The development of new housing for low-income households was much more contentious than plans for new and elaborate business districts.[‡]
- Federal programs are most accessible and directed to middle-class homeowners. Lower-income residents were only eligible for minimal compensation and had preexisting deficits in terms of housing and financial resources.
- Earthquake "ghost towns." The earthquake produced many areas of substantial commercial and residential damage. Some communities were impacted so greatly they threatened to become permanently blighted areas. This was most evident in the City of Los Angeles, especially San Fernando Valley, where more than 15% of the total available housing was destroyed. More than 9 months after the earthquake, repairs had not begun on over half of the properties in this area, and as many as one-quarter had no plans to rebuild at all. Declining property values meant many landlords had serious debts prior to the earthquake and could not afford to

---

[*] Bolin and Stanford (1998) *The Northridge Earthquake.* Routledge, London.
[†] Ibid.
[‡] Bolin, Robert, and Lois Stanford, The Northridge Earthquake: Community-Based Approaches to Unmet Recovery Needs, *Disasters* 22(1): 21, 1998. Academic Search Complete, EBSCOhost (accessed September 14, 2013).

rebuild. Others applied for loans and were still waiting, while still others were just planning to let the properties go into foreclosure.[*]
- More than 80% of damaged residential units were in multifamily housing, and low-cost rental housing was particularly affected. The bulk of the available recovery programs are designed to best serve middle-class owners of single-family homes.[†]
- Budget constraints in California at the state and local levels meant that governments were not able to fully meet matching requirements intended to ensure that federal funds are leveraged for mitigation.[‡]
- Chronic conditions of vulnerability with growing income polarization in the state, substantial migrant flows, and rapid growth of a large multiethnic low-wage working class throughout the state.

Lessons Learned

The Northridge Earthquake was one of the most costly disasters in U.S. history. While many lives were spared as a result of the date and time of the event, we can certainly foresee the increase in devastation if an earthquake of this magnitude hit this same area (or an urban area as densely populated) mid-day on an average Monday morning. In preparing communities, we must assume an increased loss of life, resulting in less ability to respond and additional services required—especially as it relates to search and rescue and hospital surge capacity. We can also assume that the impact to families would be much greater as those who contribute to the stability of a household are injured or die.

In recovery operations, we see that housing, affordable housing, and changes to the landscape in California as a result of changes to the housing market had profound impacts for recovery. In this case we know that much of the affordable housing stock was decimated as a result of the earthquake. We also know that this region suffered a deteriorating inspection capacity. Staff reductions at state and local agencies responsible for monitoring the safety of buildings and structures have resulted in the inability to vigorously enforce safety codes.[§]

The Northridge Earthquake also highlights how critical it is to be aware of immigrant populations, the unique challenges they face socially

---

[*] http://udspace.udel.edu/handle/19716/633
[†] Bolin, Robert and Stanford, Lois, *Northridge Earthquake: Vulnerability and Disaster*, 1998 Taylor and Francis, Boca Raton, FL
[‡] Ibid.
[§] Ibid.

and politically, and how that can affect response and recovery efforts. Understanding the population you serve and the policy successes and failures is of utmost importance. This means, understanding issues such as:

> There are persistent features of cultural marginalization as evidenced in the Fillmore City Council mandating that English is the "official language" of all meetings. In addition, political vulnerability was promoted by a successful California referendum (Prop 187) that has attempted to deny all health, education, and social services to "illegal immigrants" in state.[*]

Communities whose primary language is not English were marginalized prior to the earthquake. That marginalization manifests itself throughout every aspect of the recovery process. In addition, "many Latinos around here think the federal government can just load them up in box cars and ship them off to Mexico" was the sentiment in Ventura County.[†] So instead of government becoming an ally and a force for good post-disaster, marginalized Latinos distrusted government. This distrust is a common theme in nearly every case study we discuss. A lack of affordable housing predisaster and a lack of political will to build units post-disaster are common themes.

Local governments across the country are addressing language access and developing programs and policies aimed at connecting government to immigrant communities. That said, local, state, and federal governments have not done enough to understand why preexisting conditions in a community drive what happens during and after a disaster.

DISASTER EVENT: HURRICANE KATRINA (2005)

What Happened?

On Tuesday, August 23, 2005, the national hurricane center in Florida notified the public about a tropical storm system moving into the area. On August 24 the tropical storm gained strength and was named Katrina. This was the 11th named storm of 2005. By the August 25, winds had continued to strengthen into a hurricane. Its strongest winds were at about 75 mph, making it a category 1 hurricane. On Thursday around 5:00 p.m. the storm

---

[*] Bolin, Robert, and Lois Stanford, The Northridge Earthquake: Community-Based Approaches to Unmet Recovery Needs, *Disasters* 22(1): 21, 1998. Academic Search Complete, EBSCOhost (accessed September 14, 2013).
[†] Ibid.

was about 15 miles from Fort Lauderdale and about to make landfall. At 7:00 p.m. the eye of Hurricane Katrina hit Florida's southeastern coast with winds topping 80 mph. On Friday morning, August 26, Katrina began to strengthen as it crossed the very warm Gulf Coast waters. Louisiana and Mississippi declared states of emergency. On Saturday morning, August 27, Katrina's winds topped 115 mph, making it a category 3 hurricane, and by Sunday morning, August 28, winds increased again to 145 mph, making it a category 4 storm. At this point the storm was about 310 miles south of the mouth of the Mississippi River. Mayor Ray Nagin of New Orleans issued a mandatory evacuation order as tens of thousands of residents began leaving the city. Later in the day, the storm strengthened yet again, with winds of about 175 mph, making it a category 5 storm.

On Monday, August 29, at approximately 2:00 a.m., Hurricane Katrina turned toward the Louisiana coast. Throughout the morning hours Katrina began to lose force as its eye was about to come ashore. By 8:00 a.m. Mayor Nagin reported water flowing over one of New Orleans' levees. Around 11:00 a.m. the hurricane's eye came ashore, again near the Louisiana-Mississippi border—the strongest winds were about 125 mph. The strongest winds and peak storm surge slammed into Biloxi and Gulfport, Mississippi, destroying both cities. At the same time, a major levee in New Orleans failed and water poured into the city. Hurricane Katrina continued to weaken as it moved further inland; by 3:00 p.m. winds were down to about 95 mph. On Tuesday, August 30, the National Hurricane Center issued its last storm advisory as the storm that was Hurricane Katrina had winds of only 35 mph. Floodwaters continued to pour into New Orleans from failing levees throughout the city.

On Wednesday, August 31, a public health emergency was declared by the Department of Health and Human Services for Louisiana, Mississippi, Alabama, and Florida. In addition, the Louisiana governor ordered that all remaining residents leave New Orleans, but no transportation was available to operationalize the order. By September 1, the New Orleans mayor issued a plea for assistance from the federal government.[*]

Impact

Hurricane Katrina was the costliest hurricane in U.S. history and among the worst disasters this country has ever experienced (Figure 2.5). The majority of damage was due to levee failures that resulted in covering 80%

---

[*] www.news.nationalgeographic.com/news/2005/09/0914_050914_katrina

Figure 2.5 New Orleans, Louisiana, September 7, 2005. Neighborhoods and highways throughout the area were flooded as a result of Hurricane Katrina. (Photo by Jocelyn Augustino/FEMA.)

of New Orleans with up to 20 feet of water. As a direct or indirect cause of Katrina, 1,833 people were killed in five states (1,577 in Louisiana, 238 in Mississippi, 14 in Florida, 2 in Georgia, 2 in Alabama). Approximately 90,000 square miles of land area were damaged, with over 300,000 homes destroyed (10 times the number that had been similarly affected by Hurricane Andrew in 1992). The estimated economic loss was approximately $150 billion (in 2005 dollars), making it almost twice the economic loss of the 9/11 terror attacks. Total property damage was estimated at over $80 billion (2005 dollars). Over 8 million gallons of oil were spilled into the Gulf as a result of Katrina. Two hundred bodies still go unclaimed or unidentified. In the month following the storm, unemployment in Louisiana doubled, and 1 year after the storm, the population of the New Orleans region was cut in half. The forestry industry in Mississippi was also affected, with 1.3 million acres of forestlands totally destroyed, causing up to $5 billion in losses to the forestry industry. Hundreds of thousands of local residents were also left unemployed, which had a trickle-down effect in lower taxes paid to local governments.[*]

Challenges to Recovery

While mandatory evacuation orders were given in New Orleans, many people did not evacuate. For most, this was an inability to self-evacuate. Reasons cited include (especially from the poor and elderly) lack of transportation, lack of financial resources to use transportation, no place to evacuate to, fear to leave one's home, no assistance in evacuation, the feeling of needing to protect one's home. Many elderly had become ill the previous year when attempting to evacuate and then got stuck in traffic. Or, deep mistrust in government, and thus mistrust in the evacuation orders. Some did not evacuate as they had lived through many hurricanes prior and had no reason to believe that this storm would prove any different. For all the reasons cited above and more, many people died when their homes filled with water or when they were forced to their rooftops to plead for days for rescue (Figure 2.6).

Many of those who were able to evacuate also got stuck (Figure 2.7). Some were stranded on bridges attempting to leave town, while tens of thousands of others lived in squalid conditions inside the Superdome and convention center. As people died in their homes and on their rooftops, so

---

[*] For additional background and technical information on this section, please view source sites: Hurricane Katrina History and Numbers (Infographic) online at: http://www.livescience.com/11148-hurricane-katrina-history-numbers-infographic.html.

Figure 2.6   New Orleans, Louisiana, August 30, 2005. A New Orleans neighborhood inundated with floodwaters to the eaves of homes. The roof was marked by a FEMA Urban Search and Rescue (USR) team telling authorities that the structure has been searched for people. (Photo by Jocelyn Augustino/FEMA.)

Figure 2.7 New Orleans, Louisiana, August 31, 2005. Members of the FEMA Urban Search and Rescue Task Forces helped residents impacted by Hurricane Katrina. These residents were transported to the area from various neighborhoods and needed to cross over the tracks to get on a second boat, which brought them to dry land. (Photo by Jocelyn Augustino/FEMA.)

too did some who waited on bridges to exit the city or in the Superdome for aid (Figure 2.8). While the military eventually showed up to lead the local relief effort and assist in evacuation, due to inadequate communication and a disjointed government response, response operations were too late for many and recovery was prolonged indefinitely (Figure 2.9).

Due to the destruction of infrastructure, many methods of communication were rendered useless, and thus the coordination and communication on what was happening were disrupted. Civil disturbances such as looting, violence, and other criminal behavior also complicated response and recovery. Hysteria gripped the city as stories of rape at the Superdome and sniper fire targeted at rescuers spread. While there were verifiable instances of criminal activity, it was later discovered that both hysteria and deliberate exaggeration were used to bring aid to the city as quickly as possible.

The massive displacement of Louisiana and Mississippi residents to areas far outside their homes meant recovery operations had to take place locally and throughout the United States. Some cities like Chicago had

Figure 2.8 New Orleans, Louisiana, August 28, 2005. Residents brought their belongings and lined up to get into the Superdome, which had been opened as a hurricane shelter in advance of Hurricane Katrina. While many residents evacuated the city, those left behind either had disabilities or otherwise did not have transportation means to leave. (Photo by Marty Bahamonde/FEMA.)

41

Figure 2.9 New Orleans, September 5, 2005. Several boats, part of a large-scale search and rescue mission into the flooded regions of the city, departed at dawn. Individuals trapped in the city by Hurricane Katrina's floodwaters were being rescued by boat and helicopter. (Photo by Win Henderson/FEMA.)

around 10,000 evacuees, while places closer to the south, like Houston and Atlanta, had tens to hundreds of thousands (Figure 2.10). Because displacement was indefinite, the loss of people in the affected area meant less local capacity for rebuilding, as well as fewer local tax dollars, and welcoming cities were stressed with a great influx of new residents with extended needs. As people were moved around the country, separation from family members and the loss of important social networks had profound impacts on people. These impacts permeated every aspect of an evacuee's life (Figure 2.11).

Resettling in new cities without social networks and family scattered across the country meant an extraordinarily difficult platform from which to rebuild and recover. In some cases, family members were separated for many months, not knowing where their family members were, how to get in touch with them, or even if they were alive. The operations to evacuate people were so uncoordinated that people were not tracked, tearing families apart, ultimately causing extraordinary stress, fear, and instability to family members spread across the country. Reuniting family members became an important recovery operation, a significant component of disaster operations that we had not experienced in the United States

Figure 2.10 Houston, Texas, September 1, 2005. Thousands of Hurricane Katrina survivors from New Orleans were bused to refuge at a Red Cross shelter in the Houston Astrodome. (Photo by Andrea Booher/FEMA.)

Figure 2.11 Houston, Texas, September 3, 2005. An evacuee in the Houston Astrodome looks through the yellow pages for housing information. Thousands of displaced citizens were moved to Houston as New Orleans was evacuated. (Photo by Ed Edahl/FEMA.)

43

previously. Some family members were separated for such long periods of time they had to begin to resettle in the areas they were evacuated to. Given the financial strain, in the cases where family members were able to find each other, they were not then always equipped to move to be reunited with family.

Lessons Learned

Hurricanes of Katrina-like force have made landfall in the United States many times before 2005 and since. However, damages and losses of the magnitude experienced after this storm are unprecedented. Failures in communication, government coordination at local, state, and federal levels, outdated infrastructure (specifically levees), evacuation procedures, government trust, understanding of community demographics and access to resources, and responsible preparedness measures all contributed to the devastating aftermath of Hurricane Katrina. It has often been said that the impact of this storm was not the result of a natural disaster, but caused by human error and neglect. This sentiment is most certainly attributed to all the failures experienced in improperly preparing for and responding to community challenges, pre- and post-storm. Even 8 years later, we continue to see a city with only a 40% return rate and many areas still not rebuilt. Recovery is still ongoing, and many continued to live in FEMA trailers for years (Figure 2.12).

As a result of the many lessons learned post-Katrina, important discussions and actions have taken place locally and nationwide to attend to the many challenges uncovered by the storm and its aftermath. One such action has been implementation of the Post-Katrina Emergency Management Reform Act of 2006. The purpose of the act is to address various shortcomings identified in the preparation for and response to Hurricane Katrina. The act enhances FEMA's responsibilities and its autonomy within the Department of Homeland Security, of which FEMA became a part of in 2002 with the Homeland Security Act. It also amended the Stafford Act to authorize the president, in a major disaster, to call for precautionary evacuations and provide accelerated federal support in the absence of a specific request, and expanded assistance to state and local governments in recovery. The act includes several hundred additional discrete provisions as well.

Also as a federal response, the Regional Catastrophic Preparedness Grant Program (RCPGP) has been introduced to metropolitan areas across the country. According to FEMA, the purpose of the RCPGP is to enhance

Figure 2.12   In the wake of Katrina, even a decade later, residents who do return still struggle to obtain housing, put food on the table, and make ends meet.

catastrophic incident preparedness in selected high-risk, high-consequence urban areas and their surrounding regions. RCPGP is intended to support coordination of regional all-hazard planning for catastrophic events, including the development of integrated planning communities, plans, protocols, and procedures to manage a catastrophic event such as Katrina. The results of RCPGP will be made available throughout the country to enhance national resilience.

Americans have been forced to rethink vulnerabilities and risk assumptions, and that has served as a catalyst for a significant rethinking and reconsideration of how we do all disaster work. However, as we come to the 10-year anniversary of Hurricane Katrina, there continues to be a fundamental misunderstanding of how and why this storm was able to cause destruction at such an incredible magnitude. And while the legislative changes and money funneled to do preparedness planning are critical work, emergency management continues to suffer many of the same challenges and fatal errors because we have not incorporated necessary

Figure 2.13 Signs, literal and figurative, that New Orleans is still not fully recovered from the devastation of Katrina.

frameworks for understanding the preexisting conditions and unique needs in different communities (Figure 2.13).*

DISASTER EVENT: GULF COAST EVACUATION TO CHICAGO (2005–2006)

What Happened?

Given the size, scope, and unprecedented devastation caused in Louisiana and Mississippi by the Gulf Coast storms of 2005, people evacuated to cities across the country. Both the State of Illinois and the City of Chicago

---

* For additional background and technical information of this section, please view source sites: Government Accountability Office—Actions Taken to Implement Post-Katrina Emergency Management Reform Act of 2006—GAO-09-59R Published November 21, 2008 Publicly Released: December 8, 2008 Available at: http://www.gao.gov/products/GAO-09-59R

welcomed evacuees to receive recovery services. Over 10,000 evacuees entered Illinois, the vast majority to Chicago and the surrounding metropolitan area. Illinois received the largest number of evacuees of any state outside of the South (some estimates top 12,000).

The state opened mass shelter operations, and the city created a welcome center for evacuees. The welcome center was an efficient and highly effective coordinated response to all services evacuees would need (public and private) during their stay in the area (e.g., school enrollment, public transportation, public aid, social security, FEMA applications, child care). The center was a collaborative, public-private, one-stop center for accessing services. The center was open 24 hours a day, 7 days a week, and had free transportation between the shelters and welcome centers.

Impact

Around the end of September, beginning of October, it became abundantly clear that evacuees would be unable to return to their homes for an indefinite period of time. What started as an immediate recovery operation was now becoming a long-term recovery and resettlement operation. This was happening in Chicagoland, throughout Illinois, and in many of the other cities that evacuees went to (e.g., Houston, Atlanta, Denver).

Challenges to Recovery

For what might have been the first time, we see formal long-term recovery operations taking place in areas outside of the impacted area, in other cities, states, and in some cases, hundreds to thousands of miles away. This is a challenge in terms of funding and capacity. Very simply speaking, from a funding perspective, disasters receive a federal declaration and then money flows to the impacted area. In this case, money needed to get to an area completely unaffected by the event. This also requires a local political will to assist, which can be a challenge when there are other local competing demands. In the case of Chicago, much was done from the public and private sectors, not knowing for certain if, when, or how much reimbursement would be available. Over time, this can cause competition for services, and ultimately cause some local resentment from long-standing residents.

When it became apparent that a long-term recovery and resettlement operation was needed, the public and private sectors in Chicago partnered to develop a sustainable housing and employment operation. The City of Chicago's human service and emergency management departments worked with local NGOs, experts in housing, employment, and resettlement, to successfully lead the local resettlement of over 7,000 people.

The lead recovery NGO in Chicago, Heartland Alliance, focused on emergency rental assistance, assistance in expediting the FEMA process, negotiation with landlords, assistance in transferring housing vouchers, and other critical aid. Approximately 65% of the families evacuated to Chicago were at imminent risk of homelessness, and becoming homeless was prevented for all families that entered the recovery/resettlement system. According to Heartland Alliance, however, while the emergency recovery phase was successfully completed, long-term recovery needs were ever present. The most profound challenges to long-term recovery resulted from being dislocated. Like refugees worldwide, the loss of strong social networks and the adjustment to an entirely different labor market causes major challenges to sustainable recovery.

For example, evacuees who lived in the South were often able to sustain themselves through periodic work. This kind of employment is much more difficult to obtain in Chicago and is almost impossible to be enough income to survive on given the cost of living. In addition, people often relied on social networks to find employment. These networks were not as strong or were completely unavailable to evacuees resettling in Chicago. Relocation left many without critical relationships—relationships that people relied on for employment or to support employment. For example, many families shared child care with neighbors, but due to dislocation had to find and pay for formal child care.

Also, many southern states are "right to work" states. That is, people can find work even if they are not part of a union. Licensure in Louisiana and Mississippi was also not as stringent or transferable to Illinois. Many professionals, like carpenters, who owned their own businesses in New Orleans, all of the sudden could not even work in their trade due to lack of union membership or difficult-to-obtain licensures.

Finally, the labor market in Chicago is very different from those in the South. Many evacuees had jobs related to ocean industries like fishing, shrimping, large ships, etc. Those who earned a living for decades fishing faced major challenges in translating that skill to available employment in Chicago.

Lessons Learned

As disasters happen with more frequency and severity, long-term displacement and relocation becomes a much more likely result of disastrous events. We learned from the evacuation and resettlement efforts all over the country post-Katrina that loss of important social networks and immediate change in labor markets can have devastating effects on recovery due to employment challenges. In Chicago, we saw a local NGO, Heartland Alliance, expert in refugee resettlement and supportive housing, be forward thinking and develop a transitional jobs program to address the need. This allowed people to receive paid work experience, while learning a new skill or becoming familiar with the local labor market. In future evacuations, it will be critical to pay attention to where people relocate, and if there is a great difference in needed skills or little availability for transferable skills, jobs programs should immediately be developed in these cases. Attention to transfer of licensure, and perhaps assistance with that process, could also mitigate negative employment consequences due to displacement.

The successful evacuation and subsequent recovery/resettlement operations of 10,000 evacuees to Illinois illustrates the need for a quick, creative, public-private collaborative response. Partnerships and leveraging local capacities were absolutely critical to Chicago's success. However, considerable financial risk was born by nonprofits and the city that committed to serving evacuees. When the Gulf Coast storms of 2005 hit, and evacuees were relocated all over the United States, the economy was much more stable. And so, while Chicago learned much about its robust recovery ability, the exact same operation just 5 years later might not have been possible. Many NGOs have seen their budgets slashed, private giving has plummeted, many smaller nonprofits have gone completely out of business, and the City of Chicago faces another year in massive deficit, as does the State of Illinois. Therefore, in planning and responding, the immediate lessons learned (e.g., regarding labor markets and transferable skills, loss of social networks, public-private coalition building) must be coupled with jurisdictional competence to know not just what has been done in the past, but how much of it will be useful today, in the current environment. In this case, due to budget cuts and a hurting economy, a recovery operation will likely look much different even if the same exact series of events happened again.

For example, while a state might be prepared to open its doors to evacuees, it must consider if the majority of those people will flock to one area. If that area happens to be a city where public services are already stressed, an immediate influx might overwhelm a system, thereby causing those local residents already receiving services to be negatively impacted. This could then cause a secondary disaster. In addition, should a city look to its nonprofit partners, those that led operations in the past, it is important to note that those organizations, like their government partners, are also experiencing difficulties financially; then all the systems traditionally looked to for recovery are set up for a failure in proficient service delivery. It's also important to mention that while nonprofits are major leaders in disaster recovery operations, very few are compensated for this work or are able to raise money preemptively, leaving little ongoing measurable capacity. Unfortunately, much of what can be done in relief has to be determined post-event.

As Chicagoland and other major metropolitan areas work on regional catastrophic planning operations, much of which is planning for large-scale displacement, 2005–2006 post-Katrina recovery lessons learned are used as planning tools. And, as we experience the effects of nearby international disasters, such as the earthquake in Haiti in 2010, and the potential for relocation as repatriation efforts, we call into action many of the same types of lessons learned. Again, however, capacity for a comprehensive response changes quickly, and in order to see the greatest rate of success in future operations, lessons learned must be coupled with social intelligence. Lessons learned from this extraordinary recovery and resettlement operation could be misleading or even damaging, utilized on their own. Coupled with social intelligence, however, lessons learned arm emergency managers with real-time operational tools for the most effective and efficient life-saving, resilience-building apparatus.

In sum, we can prevent loss of life in the future and make communities more resilient if social intelligence is implemented in communities across the world.

DISASTER EVENT: GREAT MISSISSIPPI FLOOD OF 1927

What Happened?

In the spring of 1927, the United States experienced the most devastating river flood in the history of the country. Rains in the upper Midwestern states began in the late summer of 1926. By the beginning of September three storm systems had moved across the Mississippi Valley causing dozens of streams and rivers, from Iowa to Illinois, to overflow their banks from weeks of rain. Rain continued through September and into October, causing the Mississippi River to rise rapidly. Bridges and railroads across states were completely washed out. Rains stopped temporarily at the end of October, but by mid-December 1926, storms began again across the Mississippi Valley. Snow covered the northern states and rain fell to the south and east. By the end of 1926 people were already becoming homeless due to the flooding, and railroad traffic across the Mississippi River was suspended. Rivers across Ohio, Missouri, and Mississippi were recorded at their highest all-time levels. The gauge at Vicksburg along the Mississippi was 40 feet over its normal level.[*]

Levees were constructed along the Mississippi for 1,100 miles—from Cairo, Illinois, to the Gulf of Mexico in New Orleans. The purpose of the levees was to protect the surrounding areas from floodwaters. But by January 1927 the river had reached flood stage in Cairo, Illinois—the earliest this has ever happened. The Mississippi River remained at flood stage for over 150 consecutive days. Storms continued throughout the region for months, and by mid-April a levee just south of Cairo collapsed. Hundreds of thousands of acres of land were immediately flooded. The first levee to burst was in Dorena, Missouri, where 1,200 feet of levee crumbled. Just 5 days later massive crevices opened in levees in Mississippi and Arkansas as hundreds of millions of gallons of water violently washed over the area, drowning people and homes. Sandbagging efforts along the Mississippi commenced, but additional breaches were inevitable. Levees all along the Mississippi River collapsed or were breached as over 1 million acres of land across the Mississippi Delta were flooded by over 10 feet of water, up to 30 in some places. As of 2007, a lake in Greenville, Mississippi, still stands as a result of the flood. Near the levee break at Mounds Landing, Mississippi, the flood left a 65-acre lake that remains to this day.[†] The day of the Mounds Landing levee break, "a wall of water three-quarters of a mile across and more than 100 feet high—later its depth was estimated at as much as 130 feet—raged into the Delta . . . more water than the entire upper Mississippi ever carried, including in 1993."[‡] Throughout April and May flooding continued as levees were overwhelmed; the final breach happened on May 24, 1927, at McCrea, Louisiana.[§]

Officials were concerned about floodwaters overflowing the City of New Orleans, so on April 29, 1927, the levee at Caernarvon, Louisiana, was dynamited. The hope was to increase the speed of the river as it passed New Orleans, and therefore reduce the height of the flood wave as it passed the city. The rupture flooded the poorest sections of the city, such as St. Bernard and Plaquemines parishes. The homes and fields of the poor

---

[*] *Risk Management Solutions* 2007.
[†] Kosar, Kevin R., 2005, p. 6.
[‡] Barry, 1997, pp. 202–203.
[§] *Risk Management Solutions* 2007.

living in these areas were sacrificed for what was considered, at the time, the greater good. It turns out the destruction of the levee and the intentional flooding of one area to save another was not necessary, as breaches upstream from the city directed water away from New Orleans and into the sea along the Atchafalaya Channel. Reparations were promised to the residents of these areas from state and city officials as well as businessmen, but little payment was ever made.*

It took until mid-August for water in all the affected states to recede. The landscape was barren and totally covered in mud.

Impact

Seven states along the Mississippi River were impacted by the great Mississippi Flood of 1927: Arkansas, Illinois, Kentucky, Louisiana, Mississippi, Missouri, and Tennessee. Approximately 27,000 square miles was left underwater.[†] This area was inhabited by about 930,000 people, creating over 700,000 homeless refugees. A total of 330,000 people had to be rescued from rooftops and other high grounds. It ruined crops, damaged or destroyed 137,000 buildings, flooded 162,017 homes, killed at least 250 people (accounts vary), and destroyed over $100 million (about $1.12 billion in 2005 dollars) in crops and farm animals.[‡§] The severe flooding of fields also meant new crops could not be planted for a long time to come. The drowning of thousands of wild animals was another terrible effect of the flooding, which also had significant impacts for sectors of the economy that relied on hunting and trapping for their industry.[¶] The damage was vast and extraordinary, unlike anything the United States had ever seen. While the river flooded with some regularity, this was the first time people died as a result, and the impacts were devastating.

Direct economic losses, as estimated by the Red Cross, were $246 million ($2.75 billion in 2005 dollars). The U.S. Weather Bureau put direct losses at $355,147,000 million ($3.97 billion in 2005 dollars). Unofficial but authoritative estimates exceeded $500 million ($5.59 billion); with indirect

---

* *Risk Management Solutions* 2007.
† Ibid.
‡ Ibid.
§ Kosar, Kevin R., 2005, p. 4.
¶ The Mississipi Valley Flood Dissaster of 1927 official report of the relief operations by John Barton Payne and the American National Red Cross—Published by American National Red Cross in 1929. ARC 272. Oct. 1929

losses, the number approached $1 billion ($11.18 billion), large enough in 1927 to affect the national economy.[*][†]

Challenges to Recovery

Public utilities suffered major losses, as did businesses generally. This meant the virtual suspension of normal business that extended major economic losses beyond the immediate financial losses. Railroads, local, state, and national government property was destroyed. This had major implications for tax burdens government would have to impose in the future.[‡]

Additionally, hundreds of schools and libraries were closed, which ceased any normal educational activities for months on end. Nearly 1 million people were significantly and immediately impacted by the flood, and 325,554 people were forced to live in Red Cross refugee camps for significant periods of time, all of whom lost all worldly possessions.[§]

Racism and segregation were ever present. Many white farmers were not willing to allow black laborers to evacuate for fear that all the labor would head north. So while whites were evacuating, blacks were forced into refugee camps (in some instances atop levees as water rose), forced to work, and some were required to reinforce levees by gunpoint.[¶] Blacks comprised 75% of the population in the delta lowlands and represented 95% of the labor on the plantations and farms (as tenants, sharecroppers, and small owners). Ninety-four percent of those who were forced to flee lived in three states—Arkansas, Mississippi, and Louisiana, and 69% of those who occupied the "concentration camps" were black. Discrimination and abuse in the camps were rampant, and word was traveling fast.[**] There were too few tents, not enough food, no eating utensils in the mess halls, and black men were not allowed to leave. Those who tried were forced back at gunpoint by the National Guard. The food given to the blacks was inferior to that given

---

[*] Barry, 1997, p. 286.
[†] Kosar, Kevin R., 2005, p. 7.
[‡] *The Mississipi Valley Flood Dissaster of 1927.* Official report of the relief operations by John Barton Payne and the American National Red Cross. Published by American National Red Cross in 1929. ARC 272. Oct. 1929
[§] Ibid.
[¶] American Experience: New Orleans located at www.pbs.org/wgbh/amex/neworleans/peopleevents/p_butler.html
[**] The final report of the Colored Advisory Commission appointed to cooperate with the American National Red Cross and the President's Committee on relief work in the Mississippi Valley Flood disaster of 1927. Author: American National Red Cross—Colored Advisory Commission—Washington DC 1929.

to the whites for fear the better food would "spoil them." One white man explained that giving canned peaches to the blacks "would simply teach them a lot of expensive habits." Barry describes the camps not as refugee camps, or even labor camps, but like slave camps.* The *Chicago Defender* was reporting on the happenings, and those who could eventually left for Chicago, starting a new wave of the Great Migration.[†]

Additionally, conditions for the poor, white, and black were horrendous. Public health issues were also a serious concern. While donations poured in from tens of millions of Americans, Hollywood stars, and other countries, refugees leaving camps were totally destitute. Some received food and seed for a garden. It was estimated that a family of four would need $77.42 to clothe everyone and buy some furniture, but most left with only $27. And this was all from the Red Cross. The government itself would do nothing to help flood victims recover. The U.S. Treasury collected a record surplus of $635 million in 1927 but would not even create a loan guarantee program. Direct aid was considered charity, and charity supposedly stigmatized recipients.[‡] John Parker, governor of Tennessee, refused all outside help, even from the Red Cross, as he felt local communities "should be expected to provide for themselves rather than depend on outside assistance."[§]

The federal government's response to the disaster was a mixture of pre-New Deal minimalist government and "governing by network."[¶] The federal government made no immediate appropriations to affected areas, deciding instead to use not-for-profit organizations and a coordination of federal, state, and private resources. In short, President Coolidge directed relief via the Red Cross. In 1927 there was no federal disaster response agency, and the Red Cross was considered a quasi-government entity during times of disaster.[**]

Images of suffering continued as governors and mayors and other officials in affected states begged for President Coolidge to visit, but he did nothing and refused to visit the disaster areas. He appointed Secretary of Commerce Herbert Hoover to chair a special committee that handled the

---

* National Geogrphic "Great Flood Expedition Journal" Stephen Ambrose. May 1, 2001.
† American Experience: New Orleans located at www.pbs.org/wgbh/amex/neworleans/ peopleevents/p_butler.html
‡ Barry, 1997, p. 371.
§ Barry, 1997, pp. 371–372.
¶ Kosar, Kevin R., 2005, p. 7.
** Kosar, Kevin R., 2005, p. 7.

emergency.[*] He became one of the only heroes to emerge from this crisis and was soon referred to as "the great humanitarian" for his managing of the relief efforts. Hoover was able to create an "administrative machine" helping to clarify to the public and local, state, and federal officials who was in charge.

The flood had many consequences. "It shifted perceptions and responsibility of the federal government—calling for a great expansion." It altered politics locally and nationwide, helping Hoover sail to the presidency in 1928 and bringing a new populist governor Huey Long, to Louisiana.[††]

Lessons Learned

One of the most important legacies of the 1927 flood was the change in the role of the federal government in managing flood risk. The Flood Control Act was passed into law in 1928.[§] Prior to the Flood Control Act the United States utilized a levees-only policy of flood control.

The levees-only policy of flood control held to by the U.S. government (possibly for ease of interstate commerce purposes and increasing development opportunities) and the U.S. Army Corps of Engineers was flawed. The hotly debated levees-only policy was not in place solely to control floodwaters by damming the banks but also to increase the velocity of the river flow so that the water would scour its own bottom, essentially digging out its own channel. While this might work on some rivers, the Mississippi River is an entirely different animal. Some engineers of the day did not support the levees-only policy because they believed a more sound system was needed to create a system of alternative outlets for floodwaters.[¶] Unfortunately, many in positions of power at the U.S. Army Corps of Engineers were not scientists and in a military structure that required compliance, not contradiction, with U.S. government policies.[**] It seems there was also a strong desire from many, including the Mississippi

---

[*] National Geogrphic "Great Flood Expedition Journal" Stephen Ambrose. May 1, 2001.

[†] Barry, 1997, pp. 371–372.

[‡] Help! Call the White House! How the 1927 Mississippi Flood created big government. By David Greenberg, September 5, 2006. This article provides additional historical and technical background on the section.

[§] After the Flood: A History of the 1928 Flood Control Act by Matthew T. Pearcy published in the Journal of the Illinois State Historical Society (1998) Vol. 95 No.2 Summer of 2002 pp.172–201 Published by the University of Illinois Press.

[¶] Barry, 1997, pp. 90–92.

[**] Barry, 1997, pp. 157–159.

River Commission, to create an engineering feat, showing that man could beat the river. When levees failed, they attributed it to poor construction of the levee, not a poor levee policy. There was also little desire to spend money on the safety mechanisms. Ultimately, the lobby against creating spillways was too strong. While we no longer have a levees-only policy today, engineering of the Mississippi River is still hotly debated. We must be careful never to overestimate our ability to determine the flow of the river and underestimate the power of the river itself.

Additionally, while the flood of 1927 came as a result of unusually heavy rain for many consecutive months, the rain alone was not the sole cause of the flooding. The heavy rain, coupled with the heavy environmental impact from human impacts, left the river and rainwater with nowhere to go. For years, loggers cut down the forests above the tributaries and along the main river channel. Farmers also cleared the land for farms. This left water no place to pause before running down into the Gulf, and with no trees or roots or plants to suck up the water as in the past. So, when the heavy rains hit the valley, all the water rushed directly into the streams and tributaries filling the Mississippi River. There were no alternate routes, no breaks in the levee systems, no reservoirs, and no longer enough trees to help mitigate the impact. All the water had to move to the Gulf, and the more it rained, the more it filled and the faster it moved, causing levees to rupture and areas along the levees to experience massive flooding.*

The 1928 Flood Control Act addressed the levee-only policy through the implementation of floodways, levee channel improvement and stabilization, and tributary basin improvements. And, importantly, the legislation changed the role of the federal government in managing flood risk in the United States, as it committed the federal government, not state or local authorities, to the responsibility of flood control from Missouri to the Gulf of Mexico. At the time this was the single largest public expenditure in U.S. history, apart from World War I. A total of $325 million was appropriated for new construction, which allowed the flow of the Mississippi to be diverted in multiple locations.† The original scope of the flood control plan was even more comprehensive, including expenditures of $100 million a year to create an extensive system of reservoirs and tributary improvements. However, the cost was not politically palatable, even in the aftermath of the

---

* *Deep'n as it Come: The 1927 Mississippi River Flood* by Pete Daniel Published by the University of Arkansas Press July 1, 1998.
† *Risk Management Solutions* 2007, p. 8.

flooding. Many southerners had also called for more civilian participation in the development and implementation of any plan, but President Coolidge ignored or refused the appeals. The 1928 Flood Control Act was a major step but "gave evidence to the reality that politicians rather than engineers drove the formation of the federal flood control policy of the United States" (Pearcy, p. 190). It was a time of fiscal conservatism that precluded all the proposed infrastructure investments hoped for.

The legislation also did not include any funds to compensate the victims of the flood. A glaring weakness of the disaster relief effort was its sole reliance on private charities, particularly the Red Cross. Even in a year of a great budget surplus and public pressure for funds to be used for flood relief, the federal government resisted. Even loan programs for recovery were not offered, so those who could moved to other places, and almost all were destitute. African Americans and the poor were the most profoundly impacted. Charity was seen as shameful, even to those who were purposefully put in harm's way and promised compensation. Recovery efforts solely in the hands of private nonprofit entities will always prove to fail those most in need. We cannot relegate recovery to nonprofits, especially without the support of government funds. This did not come without consequences, however, as the lack of humanitarian aid sent shock waves across the country, resulting in a turnover from local levels of government to the presidency.

Finally, the abject racism and abuse suffered at the hands of government officials and powerful businessmen created long-lasting sentiments of mistrust. For some, this mistrust has spanned generations of those living in the intentionally flooded areas.

In all, the flood of 1927 completely changed America—from political consequences, to the creation of a massive infrastructure project (that, due to politics, still came up short), to shining a spotlight on the downfalls of mass care and, perhaps most importantly, the permanent intergenerational memories of horrific racism, mistrust, and institutional oppression. While a mass care operation now would never resemble the mass care of then (e.g., tent cities where people are segregated, served different food, forced to stay at gunpoint), we know today that inaction from government is not an option. While NGOs and charities will always have a role in recovery (and they definitely should, as they are the best suited to deliver services), government inaction or delays in government-funded recovery programs will cost government and taxpayers more in the long term. The costs will be experienced in political upheaval, loss of commerce, loss of a tax base, migration, and distrust in government. All of these come at a great cost to communities, individuals, and institutions, both those directly impacted and throughout the country, depending on the scale of the event. Major events will cause political changes locally to nationally, and tax revenue, commerce disruption, and mass migration will cost tax-payers at every level.

DISASTER EVENT: CHICAGO HEAT WAVE OF 1995

What Happened?

A brief, yet intense heat wave had developed in the central and eastern United States in mid-July 1995. Warnings of the developing heat wave were issued by the National Weather Service several days in advance of the unusually high and persistent temperatures. The warnings were quickly broadcasted by local media outlets throughout the Chicagoland area.[*] The very high temperatures for Chicago began on July 12, 1995, jumping from 90° on July 10 and 11 to 98° on July 12, 106° on July 13, 102° on July 14, 99° on July 15, 94° on July 16, and down to 89° on July 17. Additionally, not only were the daytime temperatures very high, but the nighttime low temperatures were high, remaining in the high 70s and low 80s outside.[†] These temperatures coupled with the extremely high humidity levels persisting throughout the days and evenings provided no relief from the heat for about a week, mak-ing the air "feel tropical, as if Chicago were in Fiji or Guam."[‡] The heat index (or how hot it feels based on the combination of temperature and humid-ity) for Chicago during the heat wave hit 126°.[§] Chicago was also in the core of the heat wave that hit the central and eastern United States in mid-July 1995, and therefore suffered the worst temperatures and the most deaths. Additionally, Chicago also experienced the urban heat island effect, a phe-nomenon experienced in metropolitan areas, keeping air temperatures and humidity extremely high even in the evenings, which even on the hottest days normally offer relief to people, buildings, and the air. And, with little

---

[*] 1995 report focused attention on heat's real dangers – USA TODAY. The article can be found at: http://usatoday30.usatoday.com/weather/whdie95.htm

[†] 1995 Heat Wave in Chicago, Illinois Dr. Jim Angel, State Climatologist—Illinois State Water Survey at the Prairie Research Institute. View online at: http://www.isws.illinois.edu/atmos/statecli/general/1995chicago.htm

[‡] *Heat Wave: A Social Autopsy of Disaster in Chicago* by Eric Klinenberg. The University of Chicago Press 2002, p. 1.

[§] Ibid.

to no cloud cover and no breeze, all of these conditions combined led to one of the worst heat waves in U.S. history.[*]

Impact

The extreme conditions persisted throughout the week, but emergency situations began immediately. Brick houses and apartment buildings "baked like ovens" as temperatures indoors reached 120°, even with windows open and fans blowing. Fans mostly just moved around hot air.[†] Thousands of cars broke down in the streets, roads buckled, train rails detached from their moorings, and commuters experienced long delays. City workers were called out to water bridges over the Chicago River to keep them from locking when their plates expanded. Hundreds of thousands of people crowded the Chicago beaches, and others took to their boats in Lake Michigan. The conditions were so extreme, however, that many had to return quickly, falling ill, many needing immediate hospitalization for heat-related illnesses.[‡]

Early on in the heat wave the city began to experience scattered power outages due to the record-setting use of electricity. Up to 49,000 households lost electricity at one point, some of them for up to 2 days (Klinenberg, Dying Alone, p. 1). The lack of electricity rendered air conditioning in these households useless, but also cut off access to important weather, health, and emergency information via television and radio. Elevators in buildings stopped working, complicating the removal of sick people from their homes and keeping many, especially the elderly, from exiting sweltering buildings. Many who were seeking refuge outside began opening fire hydrants to cool down large numbers of people who were not able to flee to relief in an air-conditioned space. More than 3,000 hydrants around Chicago were opened. Unfortunately, this also caused

---

[*] Droughts and Heatwaves by Ian MacLean, Joe Weiner, and Jonathan Phillip—article can be found online: www.severe-wx.pbworks.com/w/page/15957981/Droughts_and_Heat_waves

[†] *Heat Wave: A Social Autopsy of Disaster in Chicago* by Eric Klinenberg. The University of Chicago Press 2002, p. 1.

[‡] *Heat Wave: A Social Autopsy of Disaster in Chicago* by Eric Klinenberg. The University of Chicago Press 2002, pp. 1–2.

other neighborhoods to lose water pressure, and for some buildings to lose water all together, at the same time electricity was lost.[*][+]

By July 14, about 48 hours into the heat wave, thousands of Chicagoans fell victim to heat-related illnesses. Emergency rooms and ambulances were completely overwhelmed, and hospitals began to close to new patients, leaving paramedics searching the city for open beds, and others to die in their homes as their 911 call went unanswered. By the third day of the heat wave the city's morgue had gone over its capacity by hundreds. Refrigerated trucks had to be brought in to store bodies that could not be held by the morgue.[+][§]

In 1995, there were no universal standards for determining heat-related deaths. This coupled with the mayor's challenge of the staggering death numbers coming from the chief medical examiner's office meant actual heat-related death tallies were unclear for a time. After confirmation from chief medical examiners around the country, the Centers for Disease Control, and other analyses, the total number of deaths has been confirmed to exceed 700, likely 739 people. The number, 739, is the excess death rate. This is the best measure of heat deaths, as it counts the difference between the reported deaths and the typical deaths for a given period of time (*New England Journal of Medicine*, p. 1; Klinenberg, Dying Alone, p. 3).[¶]

Challenges to Recovery

Chicago is a city that is used to extremes. The city is no stranger to high and prolonged periods of heat, just as it is accustomed to long and frigid periods of cold. In fact, these were not even the highest temperatures on record. This notorious week in July was, however, a combination of

---

[*] Kenneth E. Kunkel, Stanley A. Changnon, Beth C. Reinke, and Raymond W. Arritt, 1996: The July 1995 Heat Wave in the Midwest: A Climatic Perspective and Critical Weather Factors. Bull. Amer. Meteor. Soc., 77, 1507–1518. doi: http://dx.doi.org/10.1175/1520-0477 (1996)077<1507:TJHWIT>2.0.CO;2

[+] *Heat Wave: A Social Autopsy of Disaster in Chicago* by Eric Klinenberg. The University of Chicago Press 2002, pp. 3–6.

[‡] Kenneth E. Kunkel, Stanley A. Changnon, Beth C. Reinke, and Raymond W. Arritt, 1996: The July 1995 Heat Wave in the Midwest: A Climatic Perspective and Critical Weather Factors. Bull. Amer. Meteor. Soc., 77, 1507–1518. doi: http://dx.doi.org/10.1175/1520-0477 (1996)077<1507:TJHWIT>2.0.CO;2

[§] *Heat Wave: A Social Autopsy of Disaster in Chicago* by Eric Klinenberg. The University of Chicago Press 2002, pp. 3–6.

[¶] http://journals.ametsoc.org/doi/pdf/10.1175/1520-0477%281996%29077%3C1497%3AIAR TTH%3E2.0.CO%3B2

weather extremes (high temperatures, high dew points/humidity, and heat island effects).

Among the most notable features of this event were the very high temperatures coupled with record-breaking dew point temperatures over the southern Great Lakes region and the Upper Mississippi River Basin.* A dew point is the water-to-air saturation temperature (or the point at which dew forms), and it is associated with relative humidity. A high relative humidity means the air is saturated with water (100% relative humidity equals total saturation). This is important because the higher the humidity, the less able people are to use thermoregulation to cool themselves. When temperatures are high, people cool themselves when their perspiration evaporates (i.e., thermoregulation), but when it's very humid, the air is already too saturated with water to allow perspiration to evaporate, so humans remain very hot as they cannot cool.

Additionally, the relatively new phenomenon called an urban heat island was in effect during the heat wave. An urban heat island refers to the increase of temperature in a city in comparison to its surroundings. This is a result of buildings, roads, and other infrastructure replacing open land and vegetation. Surfaces that were once permeable and moist become impermeable and dry. And because there is not enough vegetation to absorb the sunlight, the sunlight is instead absorbed by man-made structures (roads, parking lots, buildings) that cannot evaporate water, so the sunlight's energy raises the temperature of those surfaces and the air around them. This causes a dome of warm air over a city, which persists during the night even when the sunlight is gone, forming an "island" of high temperatures. On a hot sunny day the sun can heat dry exposed urban surfaces to temperatures 50–90° hotter than the air (while areas with more vegetation will remain close to air temperatures). Urban heat islands persist throughout the evening, even when the air can cool, because infrastructure will slowly release heat. The annual mean air temperature of a city with 1 million people or more can be 1.8–5.4° warmer than its surroundings. On a clear, calm night like Chicago had, however, the temperature can be as much as 22° higher.†

All this said, the mayor did not consider this an emergency until people started dying. And even then, the administration was publically questioning how bad things really were. By the time the administration stopped try-

---

* http://journals.ametsoc.org/doi/abs/10.1175/1520-0477%281996%29077%3C1507%3ATJH WIT%3E2.0.CO%3B2

† http://www.epa.gov/heatisld/about/index.htm

ing to deny there might be a problem, it was already way behind the eight ball. Hospitals were closing, as they became completely overwhelmed; there were not enough ambulances, no real heat emergency plan existed, and the skimpy one Chicago relied on was not activated. Mayor Daley was quoted as saying this was "not a big deal." Many members of the mayor's administration were publically blaming family members for not taking care of one another. Chicago's Human Services Commissioner stated, "We are talking about people who die because they neglect themselves."[*] At the same time the city refused to call in desperately needed ambulances from the suburbs, which could have taken people to hospitals just outside the city. The police department did not use its senior units to attend to the elderly residents they are charged with protecting. Eventually, when cooling centers were opened, many still had no way to get to them. As the city spent the majority of its time and effort on a massive public relations campaign to shape public opinion rather than respond to the disaster, Chicago experienced the most deadly event in its history.[†]

While Commonwealth Edison was blamed for major power outages, many people did not have air conditioning anyway and cooled off by sleeping in the parks or in other common spaces. However, while areas of the city once experienced safe sleeping in the park, neighborhoods had changed rapidly, keeping many, especially the elderly in African American neighborhoods, in their houses due to fear for their safety. Even in the cases when city workers came to knock on doors, not opening the door was a survival strategy for living alone in the city.[‡]

Lessons Learned

On average, more people die each year in the United States from heat-related deaths than all other disasters combined. This is also a phenomenon worldwide where drought and heat waves cause more deaths than any other weather disasters combined.[§¶] That said, the loss of life

---

[*] *Heat Wave: A Social Autopsy of Disaster in Chicago* by Eric Klinenberg. The University of Chicago Press 2002.

[†] *The Case Against Daley* by Steve Rhodes published June 21, 2007 in Chicago Magazine

[‡] *Heat Wave: A Social Autopsy of Disaster in Chicago* by Eric Klinenberg. The University of Chicago Press 2002

[§] *Droughts and Heatwaves* by Ian MacLean, Joe Weiner, and Jonathan Phillip. Article can be found online: www.severe-wx.pbworks.com/w/page/15957981/Droughts_and_Heat_waves

[¶] "Dying Alone: An Interview with Eric Klinenberg," found at www.press.uchicago.edu/Misc/Chicago/443212in.html

in Chicago during the heat wave was preventable. "Given this advance warning, many, if not all, of the heat-related deaths associated with this event were preventable." As former NOAA Chief Scientist and leader of National Disaster Survey Team which investigated event—Kathryn D. Sullivan states, "So what went wrong?"

According to the report, in Chicago and Milwaukee, a heat wave of this magnitude is so unusual that it was not immediately recognized as a public health emergency. The heat wave was a highly rare—in some respects an unprecedented—weather event because of its unusually high maximum and minimum temperatures and accompanying high relative humidities. "Unfortunately, a heat wave connotes discomfort, not violence; inconvenience, not alarm," said Sullivan.

We also know that the majority of people who died were elderly and African American. Most of them died in their homes, isolated and alone. The heat disaster that took place in Chicago in 1995 is not unlike the disaster in New Orleans 10 years later. Both uncovered preexisting social problems that, when coupled with extreme weather, caused a cascade of devastating events resulting in a social disaster. So while the "forces of nature played a major role . . . these deaths were not an act of G-d."[*]

According to experts like the U.S. Centers for Disease Control and Prevention, a typical victim from the Chicago heat wave was living alone, not leaving home daily, lacking access to transportation, sick or bedridden, had no social contacts nearby, and without an air conditioner. But according to Klinenberg, this doesn't even come close to telling the entire story, as given these factors, female victims would have outnumbered male victims because women tend to live alone at higher rates as they get older. But as Klinenberg learns through his extensive research, men were twice as likely to die as women, and this has everything to do with the close social relationships elderly women retain as they age, but men tend to lose.[†]

Additionally, while death tolls for African Americans and whites were nearly the same, many more elderly whites live in Chicago than do elderly African Americans. And, when Chicago's Department of Public Health investigated the age difference, they found the black-to-white mortality ratio to be 1.5 to 1.[‡]

---

[*] "Dying Alone: An Interview with Eric Klinenberg," found at www.press.uchicago.edu/Misc/Chicago/443212in.html
[†] Ibid.
[‡] Ibid.

Klinenberg also finds another surprising fact: Latinos in Chicago represented about 25% of the population in 1995, but only accounted for 2% of the deaths, despite being disproportionately sick and low-income. Many Chicagoans and local experts talked about the presence of family values in Latino families to account for the differences in death rates during the heat wave. However, as Klinenberg found, there is a critical "social and spacial context" that must be considered when considering close family ties. That is, Chicago's Latino neighborhoods tend to have high density with central and vibrant public spaces, along with busy commercial life in the streets, while most of the African American neighborhoods that had the highest death tolls had been abandoned by many stores and employers, causing many residents to move. When a neighborhood becomes sparse, family and social connections become incredibly difficult to maintain.*

There is no reason that any human being should be so underserved and unnoticed that he or she is allowed to die alone and isolated. We learned from the 1995 Chicago heat wave that better warning systems are absolutely critical. And, these warning systems need to take into account different neighborhood dynamics and varying degrees of social systems. Chicago is known as a city of neighborhoods, and each neighborhood has its own unique strengths and challenges. You can take a 2-hour drive on any one street from the northern border of the city to its southernmost point and see this revealed. The deaths in Chicago were a result of the increased numbers of isolated seniors, living in socially vulnerable situations, often fearful, rarely contacted, and in neighborhoods with no neighbors, people, government, or businesses. We must acknowledge challenges beyond the weather event and know our populations well enough to tailor responses.

We also need to continue the conversation on legislation for cooling subsidies and subsidies for air conditioners. The Low Income Home Energy Assistance Program (LIHEAP), which is always subject to massive cuts, needs to be able to maintain sufficient funding levels, a small preventative cost on the front end that we know will ultimately save lives.

Environmental considerations that are unique to urban areas must also receive close attention when preparing for extreme weather events. As cities develop and continue to get larger, both up and out, open green space is lost. This makes our cities hotter and for longer periods of time. We must consider permeable paving, green roofs, solar and wind energy, and just more open green space, with trees and native plantings that don't

---

* "Dying Alone: An Interview with Eric Klinenberg," found at www.press.uchicago.edu/Misc/Chicago/443212in.html

require as much watering. If we can keep our cities cooler naturally, we can be much more resilient.

As Klinenberg concludes, "The heat wave was a particle accelerator for the city: It sped up and made visible the hazardous social conditions that are always present but difficult to perceive. Yes, the weather was extreme. But the deep sources of the tragedy were the everyday disaster that the city tolerates, takes for granted, or has officially forgotten."*

DISASTER EVENT: ILLINOIS FLOODS (2010)

What Happened?

On Friday, July 23, and Saturday, July 24, 2010, a severe rainstorm hit central and northern Illinois. Heavy rain continued over a 24-hour period, quickly dumping more than 7 inches into the region (Figures 2.14 and 2.15). The torrential rains temporarily forced closures of many highways, public transportation systems, and beaches, as flooding forced sewage

Figure 2.14 Hundreds gathered for sandbagging efforts in the region.

---

* "Dying Alone: An Interview with Eric Klinenberg," found at www.press.uchicago.edu/Misc/Chicago/443212in.html

Figure 2.15 National Guard, men and women, old and young alike, pulled together in preparing sandbags in efforts to combat flooding.

water to be released into the lake (Figure 2.16). By Sunday some of the rainwater began to subside, but many areas still remained under flood warnings. Crews set out in boats to rescue people trapped in the flooded homes or in cars in viaducts.*

Impact

The severe storm left over 150,000 people without power. By the end of August, seven counties in Illinois received a federal disaster declaration as a result of the flooding caused by the storm (Figure 2.17). The seven counties included Carrol, Cook, DuPage, Jo Daviess, Ogle, Stephenson, and Winnebago. Cook County, where the City of Chicago resides, was the hardest hit. As of the beginning of November 2010, FEMA had received over 137,000 applications for assistance—over 131,000 in Cook County alone. And, over $320 million has been spent in individual grants and low-interest loans to flood victims

---

* NBC Chicago July 25, 2010—Floodwaters Starting to Recede sourced online from: http://www.nbcchicago.com/weather/stories/Floodwaters-Starting-to-Recede-Chicago-flood-heavy-weekend-storms-weather-damage-July-25-2010-99193489.html

Figure 2.16   Numerous highways and bridges in the region were closed.

Figure 2.17   Hundreds of roads, houses, and businesses were completely flooded.

statewide. Most of this aid—about $307 million—went to people in Cook County. This is the largest disaster assistance package for individuals in the State of Illinois' history. As the time to apply for assistance neared its conclusion, people continued to apply for assistance every day.*

Challenges to Recovery

Recovery post-flood was a challenge, as such a large number of people who were eligible for assistance never applied, were not able to apply in time, or didn't know what the process for assistance was. The top 10 reasons people have cited for not applying are

1. "I thought I applied months ago when I reported my damages to local officials or filled out a form for my county."
2. "I thought if you had insurance (a sewer backup rider), you couldn't apply."
3. "I thought renters couldn't get assistance, just homeowners."
4. "I thought the media said assistance was denied for Cook and DuPage Counties." (That only applied to infrastructure damages for municipalities and state agencies.)
5. "I thought you could only receive a loan, and I don't want a loan."
6. "I thought you couldn't apply if you have already cleaned up the repairs."
7. "I thought any assistance would be taxable or affect my social security payments."
8. "I thought you had to live in the City of Chicago to qualify."
9. "I thought the assistance was welfare and I wouldn't qualify."
10. "I didn't know there was any assistance available for the flooding."

Misinformation and lack of information were enormous impediments to the recovery in Illinois. In addition, as it took a month for the federal declaration to get signed, tens of thousands of homes sat with substantial water for extended periods of time in hot summer months (Figure 2.18). And, to complicate matters, because the hardest-hit areas did not receive a public assistance declaration, local government departments and area nonprofits that traditionally assist in disaster recovery work received no federal assistance to shoulder the burden of extra services as a result of

---

* For further background on this section, visit: http://articles.chicagobreakingnews.com/2010-11-26/news/28519975_1_flood-victims-flood-damage-illinois-counties-disaster-areas and http://www.fema.gov/news/newsrelease.fema?id= 53221

Figure 2.18   Roscoe, Illinois, August 5, 2010. Residents in Roscoe, Illinois, clean up after torrential rain showers caused flooding in the area. FEMA worked with the Illinois Emergency Management Agency (IEMA) to gather disaster estimates to send to the governor's office. (Photo by Patsy Lynch/FEMA.)

the disaster. This is a region that is also already suffering massive budget deficits and decreased private giving, so very little extra capacity is available to help families in the recovery process.

These floods also hit poorer communities the hardest. All 10 Cook County neighborhoods and suburbs getting the most disaster aid have more families and individuals living below the poverty level than census data show for the rest of the United States. The homes of people who were poorer were more likely to flood, and they were less able to recover. Many of those in affected communities were elderly and unable to do the home cleanups and mold removal on their own.*

Lessons Learned

As the local recovery continued, the Chicagoland area learned how quickly a situation can go from bad to worse when vulnerable communities are so

---

* http://www.chicagobreakingnews.com/2010/11/fema-poorer-areas-hardest-hit-finan-cially-by-july-flood.html; http://www.fema.gov/news/newsrelease.fema?id = 53221

heavily impacted. In this case, poor and elderly households had inches to feet of water in their homes. Because there was little to no local aid available, and federal aid was slow in coming, what started as some water in basements resulted in families facing homelessness a few months later. Mold grows rapidly, and the cost of removal is significant, especially for those with little disposable income or those living paycheck to paycheck. In addition, much of the debris removal was done by people on their own—many elderly households did not have the capacity to undertake such labor.

When it became apparent that tens of thousands of people would need assistance in debris and mold removal, local faith-based groups, community groups, and nonprofits organized to bring assistance to families. However, the need was greater than the availability of volunteer assistance. In the past when floods or other events have hit the area, assistance from the heavily relied upon nonprofit community was available. However, in 2010, after 2 to 3 years of substantial cuts to nonprofits, from both public and private funders, capacity was diminished severely. And, with the City of Chicago facing another year in massive deficit ($655 million), as well as the State of Illinois ($13 billion), it is no wonder local resiliency has decreased substantially. The absence of public and private money for assistance requires preexisting community stability, and if a region is already facing massive budgetary concerns, no extra resources are available to serve the additional need— from both a private and a public perspective.

Finally, because federal officials make decisions on public assistance (i.e., federal aid to local governments and nonprofits) based on average cost per county resident, it is inherently unfair to those more populated counties, especially in a place like Cook County, the second most populous county in the United States.

DISASTER EVENT: GULF COAST OIL SPILL (2010)

What Happened?

On April 20, 2010, an underwater explosion beneath the oil rig *Deepwater Horizon* set off a chain of events that cost 11 people their lives and dumped millions of barrels of oil into the Gulf of Mexico. This man-made disaster wreaked havoc on the environment and destroyed the economy of many communities along the Gulf Coast.

Impact

Over a 2-month period, millions of barrels of crude oil seeped into the Gulf of Mexico. Emergency response included the deployment of skimming ships to pick up oil, environmental scientists with backgrounds in oil spills, drilling teams, and engineers from all backgrounds working together with the government and BP to cap the leaking rig. The leaking rig was capped on July 15, 2010, and permanently sealed on September 19, 2010. The 6-month-long disaster is one of the worst the world has seen.[*] Within 4 months, 3,090 birds and 270 sea turtles had been reported visibly oiled, and 3,765 were found dead[†] (Figure 2.19). In addition to the loss of wildlife, nearly 30,000 people responded to the disaster, along with 4,000 ships, to respond to a 3,800-square-mile disaster zone.[‡] According to *Forbes Magazine*, a February 2013 estimate pegged the total costs accrued by BP at $43 billion.

The oil spill damaged and nearly destroyed local communities and their economies. Some of the immediate impact included

- With commercial fishing accounting for one-third of the employment in the affected area, processing plants and canning plants were impacted.
- Retail and other support businesses that sprout up in single-sector communities suffered.
- Tourism and the hospitality industry suffered due to the stigma associated with the disaster.
- Retail and support businesses supporting the tourism and fishing communities suffered.
- More indirectly, the federal moratorium on deepwater drilling in the Gulf forced oil companies to abandon offshore drilling projects, and their oil rig employees had to find alternative employment.

---

[*] President Barack Obama, Remarks by the President to the Nation on the BP Oil Spill, June 15, 2010. Available at http://www.whitehouse.gov/the-press-office/remarks-presidentnation-bp-oil-spill.

[†] U.S. Fish and Wildlife Service, Department of Interior, *Deepwater Horizon Response Consolidated Fish and Wildlife Collection Report*, 2010. Available at http://www.fws.gov/home/dhoilspill/pdfs/collection_08012010.pdf.

[‡] Jarvis, Alice-Azania, BP Oil Spill: Disaster by Numbers, *Independent*, September 14, 2010. http://www.independent.co.uk/environment/bp-oil-spill-disaster-by-numbers-2078396 (reporting 4,768 as the number of dead animals collected as of August 13, 2010, of which 4,080 were birds and 525 were sea turtles).

Figure 2.19 Jefferson Parish, Louisiana, June 6, 2010. Carl Pellegrin (left) of the Louisiana Department of Wildlife and Fisheries and Tim Kimmel of the U.S. Fish and Wildlife Service prepare to net an oiled pelican in Barataria Bay, Louisiana. The pelican was successfully netted and transported to a facility on Grand Isle, Louisiana, for stabilization before being taken to Fort Jackson Oiled Wildlife Rehabilitation Center in Venice, Louisiana, for cleaning. State and federal wildlife agencies are coordinated across the Gulf Coast to rescue wildlife affected by the *Deepwater Horizon* oil spill. (Photo by Coast Guard Petty Officer 2nd Class John Miller.)

Challenges to Recovery

Later in this book, you will read more about single-sector reliant communities. That is, a single employer would create a large enough tax base to support retail, hospitality, and other businesses. The result: a community is built around the domino effect of one large employer. Any economic shock (e.g., recession, disaster, etc.) would result in a ripple effect throughout this community. For example, a 5% reduction in employment could stress local food pantry usage, threaten the social safety net, reduce tax revenues, etc. The challenge to recovery along the Gulf Coast in the corollary is that the Gulf of Mexico is the company and the entire region is the company town.

The impact of the oil spill to communities was immediate. Many Gulf Coast communities are wholly reliant on the ocean. Some communities also rely heavily on the oil and gas exploration industry as an economic engine.

Bayou La Batre is known as the "Seafood Capital" of Alabama and is dependent on resources such as shrimp, oysters, crabs, mullet, and other finfish. Gulf waters, where a high volume of these resources are typically harvested, were contaminated by the oil spill. In addition, community commerce includes shipbuilding, marine supply businesses, marine repair shops, and other businesses that cater to the commercial harvesting and processing of seafood. On the other hand, Dauphin Island relies on tourism based on beaches and beach house rentals, boating, recreational fishing, and charter boat tours, which are all tied to natural resources affected by the oil spill.[*]

The spill severely damaged and threatened many at-risk industries along the northern Gulf, including commercial and recreational fishing, tourism, and other enterprises tied to natural resources.[†] Notwithstanding the immediate impacts, long-term issues include water and air quality, dispersant use, beach contamination, tourism, and the claims process.[‡] Shorter-term issues included payments from BP to claimants, perception issues related to fishing in the Gulf Coast waters, and the impact of high unemployment on local tax bases and related municipal services. Approximately one-third of the U.S. territory in the Gulf was closed to commercial fishing, leaving many fishermen without gainful employment.[§] In addition to the immediate impacts, there were cascading impacts to communities when fishermen and oil/gas exploration workers found themselves unemployed.

While fishermen and fishing-related entities were immediately impacted, supporting industries and sectors were impacted shortly after. If the fishermen and their families were without incomes, so were the local restaurants and shops.

---

[*] Gill, Duane A., J. Steve Picou, and Liesel A. Ritchie, The *Exxon Valdez* and BP Oil Spills: A Comparison of Initial Social and Psychological Impacts, *American Behavioral Scientist* 56(1): 3–23, 2012.

[†] Ibid.

[‡] Gill, Duane A., J. Steve Picou, and Liesel A. Ritchie, The *Exxon Valdez* and BP Oil Spills: A Comparison of Initial Social and Psychological Impacts, *American Behavioral Scientist* 56(1): 3–23, 2012, p. 8.

[§] Davis, Andrew B., Pure Economic Loss Claims under the Oil Pollution Act: Combining Policy and Congressional Intent, *Columbia Journal of Law and Social Problems* 45(5): 1–44, 2011. Academic Search Complete, EBSCOhost (accessed August 13, 2013).

Many of the businesses in the supply chain are family businesses, run by generations of the same family. The ties and networking relationships can go back decades. Yet, as these are small businesses with limited resources, it is difficult to know just how well they can endure considering three consecutive extraordinary events: Hurricane Katarina in 2005, the BP oil spill in 2010, and the Mississippi River floods of 2011.[*]

The oil spill, almost overnight, immediately threatened the near- and long-term survival of communities. Because the impacts to the economy were so immediate, the public outcry was deafening. People were looking to hold BP Amoco responsible for the economic impacts resulting from the oil spill. From a crisis management perspective, BP executives bungled the response operations with mixed messages about the amount of oil seeping into the Gulf and how long it would take to cap the leaking rig. As each day and week passed, the outrage compounded.

Seeking to repair its damaged reputation, BP Amoco set aside nearly $20 billion in funds to cover Gulf Coast recovery efforts. Even with the setting aside of funds, accessing the funds was difficult and, in many cases, compounded the impacts of the disaster. Navigating the bureaucratic process to receive payments was cripplingly slow and left people without any financial relief for weeks and months. Consider the following quotes from news coverage as they capture the experiences of affected individuals and families.

- "They pay my son, and turned me down. We fish on the same boat with the same license," said one person at the meeting.[†]
- "My 11-year-old child has tested positive for the chemicals from the oil. Who is going to see him?" asked another person at the meeting.[‡]
- One after another, they described being turned down unjustly, forcing Feinberg to offer on-the-spot help.[§]
- "I got 62 pages of documentation. 62 pages," said one angry man. "Everything y'all asked for I provided. Y'all say 30 to 90 days away for a minimum claim to be paid for? What is my family supposed to do until then?"[¶]

---

[*] Lambert, John, David Duhon, and Joseph Peyrefitte, 2010 BP Oil Spill and the Systemic Construct of the Gulf Coast Shrimp Supply Chain, *Systemic Practice and Action Research* 25(3): 223–240, 2012. Academic Search Complete, EBSCOhost (accessed August 14, 2013).

[†] http://www.wwltv.com/news/Angry-Questions-For-Ken-Feinberg-in-Bay-St-Louis-113235109.html.

[‡] Ibid.

[§] Ibid.

[¶] Ibid.

- "Now they're telling me it is going to take 90 days for me to go through this interim process, which I don't have 90 days, because next week, I'm not going to have money to eat with. My lights will be turned off."[*]

Even after BP set aside billions to assist in recovery efforts, communities were facing a long-term recovery. That is, even after the oil slicks and environmental issues were remediated, impacted communities were going to face the long-term challenges.

While a recovery fund was set up to accelerate recovery, and Kenneth Feinberg was brought in to administer the recovery fund facility (Feinberg was the fund administrator for the post-9/11 fund), many politicians and policy makers along the coast found the process to access funds "opaque, arbitrary and slow."[†] While policy makers and politicians complained about the process, BP argued the recovery packages were too generous because they "overestimated" future economic losses.[‡] All said, what ensured a challenging recovery for the Gulf Coast was the dependence on the Gulf of Mexico as the economic engine in the same way a company town relies on a single large employer to sustain its tax base.

By June 2011, Fienberg distributed about $3.5 billion of the $20 billion provided to the recovery facility.[§] At that point, some 100,000 people had filed for a final settlement, with an additional 90,000 opting to take a quick-pay process that settles all claims with a payment of $5,000 to individuals and $25,000 to businesses.[¶]

A reality often lost on policy makers is the fact that many communities in 2010 were just beginning to recover from Hurricanes Katrina and Rita. For example, many fishermen and their families took out personal loans and lines of credit to repair and purchase fishing equipment. With financial obligations still outstanding, any shock to the economic system would have a catastrophic impact on families and communities. This reality was also lost on BP as it continued to profess that a $20 billion recovery facility was too generous. The painstakingly slow process to access recovery funds sent individuals into a tailspin, and the result was that there was overwhelming pressure on governments and shrinking tax bases while a simultaneously huge demand for

---

[*] http://www.academia.edu/4313624/Stretching_the_Bonds_The_Families_of_Andrew, p. 154.
[†] http://www.nytimes.com/2011/02/18/us/18bp.html?_r=0.
[‡] Ibid.
[§] Ibid.
[¶] Ibid.

services multiplied, and no easy answers. Like previous case studies, the same issues (preexisting conditions) present themselves during the recovery phase.

Lessons Learned

It appears that at least in the near term, our American demand for cheap oil and gas is unlikely to diminish. And for this reason, oil and gas exploration along domestic shores is likely to continue. As a result, future oil spills are likely—while oil companies are spending more on safety mechanisms, accidents will happen. The ultimate lesson learned after the *Deepwater Horizon* spill: the impacts from the spill and Hurricanes Katrina and Rita will be felt in the Gulf Coast region for decades. The reality that comes with an extended recovery includes understanding how people and communities will need help in the years after the disaster.

Much of the Gulf Coast relies on the Gulf of Mexico to drive the economy. Fishing, fish processing, oil and gas exploration, and tourism industries created other supporting industries like retail and hospitality. When combined, a healthy tax base emerges. Any shock to the delicate ecosystem (and specifically any disruption to industries reliant on the Gulf of Mexico), and it all comes crashing down at once. As a result of the oil spill, and the disruption to the Gulf Coast natural and social ecosystem, there were some major lessons learned.

1. With one-third of commercial fishermen out of work, and many of the fishermen already financially extended from Hurricane Katrina, financial safety nets were stressed.
2. Many commercial fishermen were Vietnamese refugees. Accessing disaster relief funds from the BP recovery facility would prove difficult to impossible due to cultural factors and language barriers.
3. Much of the seafood sourced from the Gulf Coast is harvested. The oil spill devastated many of the habitats and ecological conditions required to raise and harvest seafood.
4. The payment process was very slow, which forced many families into dire financial circumstances.
5. A federal moratorium on oil and gas exploration and drilling devastated the Gulf Coast oil and gas industry. As a result, many high-wage workers in the area suffered through extended periods of unemployment.
6. As extended unemployment drained savings and other safety nets, individuals and families triaged financial obligations. As a

result, tax revenues declined—sales tax, income tax, and property tax revenues all declined.

7. A decline in tax revenues stresses government's ability to fund capital improvements, basic services, and necessary programs.

8. Tort law does not allow people impacted by technological disasters to recover damages that may result from an oil spill. For example, the oil spill destroys the fishing industry, which severely impacts the tourism industry. Tort limitations do not allow claimants from the tourism industry to seek damages from the oil company. While BP voluntarily created a recovery facility, even BP felt disbursements connected with domino impacts were overly generous.

As we see over and over again, disasters impact various communities differently. That is, the demographic and socioeconomic makeup of a community plays a significant role in how a community is impacted by disaster. Refugee populations tied to the fishing industry were more dramatically impacted than most fishing communities because of cultural and language access barriers. Mental health issues were more prevalent among vulnerable communities, and people in these communities are often the last to seek mental health services. Perhaps most evident in this disaster is the domino effect to the economy—in addition to the environmental impacts, the oil spill resulting from the *Deepwater Horizon* explosion created a tidal wave of economic destruction. Absent any changes to tort law or to regulations surrounding oil and gas exploration, the Gulf Coast remains vulnerable to a similar incident. In addition, with the frequency and intensity of hurricanes and super-storms, the entire Gulf Coast appears to barely recover from one disaster event before being thrust into another.

As we see over and over again, vulnerable populations are more dramatically impacted than more resilient populations and communities. In this case, the entire Gulf Coast functions like a company town. And when the "company" is threatened, the entire community deals with the impacts.

DISASTER EVENT: HURRICANE SANDY (2012)

What Happened?

On October 29, 2012, Hurricane Sandy made landfall in southern New Jersey, with impacts felt across a dozen states[*] (Figure 2.20). The area

---
[*] http://www.fema.gov/hurricane-sandy-timeline.

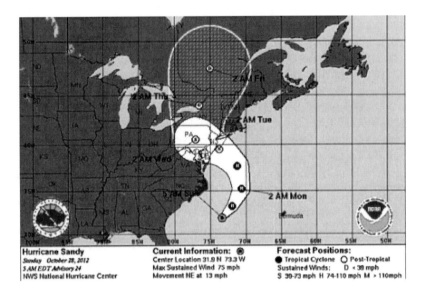

Hurricane Sandy	Current Information: ●	Forecast Positions:
Sunday October 28, 2012	Center Location 31.9 N 73.3 W	● Tropical Cyclone ○ Post-Tropical
5 AM EDT Advisory 24	Max Sustained Wind 75 mph	Sustained Winds: D < 39 mph
NWS National Hurricane Center	Movement NE at 13 mph	S 30-73 mph H 74-110 mph M > 110mph

Figure 2.20 October 28, 2012. A National Oceanic and Atmospheric Administration (NOAA) map of the possible projected storm track prior to Sandy hitting the East Coast. (From NOAA.)

of impact includes some of the most densely populated areas in North America and the world. In addition to population density, the Eastern Seaboard of the United States is home to the center of the global financial markets, transportation centers, and universities. While the storm made landfall as a category 3 hurricane, newscasters dubbed Sandy a superstorm because of sheer size. Sandy was the largest tropical cyclone in the extended best track dataset, which began in 1988.[*]

In sum, as a result of the storm, nearly 23,000 people were left homeless, over 130 lost their lives, and 8.5 million people were left without power.[†] The size of the storm and the amount of destruction it caused across the U.S. Eastern Seaboard make it one of the costliest disasters in U.S. history (Figure 2.21).

---

[*] Blake, Eric S., Todd B. Kimberlain, Robert J. Berg, John P. Cangialosi, and John L. Beven II, Tropical Cyclone Report: Hurricane Sandy (AL182012) 22–29 October 2012, National Oceanic and Atmospheric Administration, National Hurricane Center, February 12, 2013.
[†] Ibid.

Figure 2.21 Breezy Point, New York, November 14, 2012. In the wake of Hurricane Sandy, debris and destruction can be seen in and around the houses in Breezy Point, New York. Over 100 houses burned to the ground as floodwaters isolated the community from firefighters. Hurricane Sandy was the largest Atlantic hurricane on record and caused the most damage in New York and New Jersey. (Photo by U.S. Navy Chief Mass Communication Specialist Ryan J. Courtade/Released.)

Impact

The impact was immediate. As one of the biggest storms to ever make landfall, Hurricane Sandy's impacts were felt immediately. They included

- Nearly 8 miles of coastline were destroyed.
- Over 8 million people were without power after the storm.
- Over 20,000 people sought emergency shelter.
- Nearly $3.2 billion in federal aid was distributed to state and local governments.
- Over $1.4 billion in disaster assistance was provided to individuals.
- The total cost of the disaster was $65 billion— second only to Hurricane Katrina.
- Power outages and a lack of temporary housing exacerbated recovery efforts.

- Widespread flooding in low-lying areas of New York City and flooded subway tunnels made traversing the disaster-impacted areas very difficult.

Damages were vast and hit nearly every human system. The storm caused billions in damages to public transit systems, there were impacts to local tourism, and vulnerable populations were unevenly impacted.

Challenges to Recovery

At the time this book is being written, Sandy recovery is still a work in progress. The storm massively impacted human and ecological systems. For people, especially the socially vulnerable, the impacts from Sandy were felt immediately. With the number of power outages nearing 10 million, and widespread flooding, access to food, water, and medicine was difficult. In addition, the widespread and regional destruction stressed government at every level. With finite resources, some communities waited longer than others to navigate the post-storm environment. This meant that some communities were without power for days, and in some cases weeks. Impacts from power outages included

- A lack of access to basic resources in the immediate aftermath of the storm—water, food, medicine, housing, etc.
- Some seniors and the disabled relied on refrigerated medicine. Without power, they were unable to access critical medicine.
- Without access to public transit (some subway tunnels were completely flooded by storm surge), people were trapped in their homes.
- Over 20,000 New Yorkers sought emergency shelter, and according to DNAInfo New York, nearly 40,000 New Yorkers were homeless as a result of the storm.[*]
- Hundreds of families remained living in hotels months after the storm.
- Many of the affordable housing/public housing complexes were damaged by the storm.
- A general lack of affordable housing prestorm made locating new affordable housing difficult.

---

[*] http://www.dnainfo.com/new-york/20121104/new-york-city/up-40000-new-yorkers-left-homeless-after-hurricane-sandy-mayor-says.

While New York City and the Eastern Seaboard are home to many of the same social vulnerabilities described in the previous case studies, the size and magnitude of the storm was such that we didn't hear the pleas of just the poor and vulnerable. The storm impacted millions of people across a region.

All said, many of the same challenges vulnerable populations face and government struggles to understand were apparent during Sandy.

Lessons Learned

While the recovery process is still ongoing, there are some clear lessons learned from Hurricane Sandy. They include

1. The entire seaboard needs to storm-proof its shorelines. Nearly 100 miles of beach eroded during the storm. Beaches and wetlands need to be restored, and governments across all levels must work together toward more sound development principles (that account for future storms) along the coast.
2. Major cities, especially those vulnerable to storms resulting from impacts of climate change, must begin to develop policies and programs to build more affordable housing.
3. All units of government must work together to eliminate vulnerabilities to electrical grid and other power delivery infrastructure.
4. We must better understand vulnerable populations across a jurisdiction prior to disaster. With millions of people without power, developing a disaster supply chain to deliver critical resources to vulnerable populations is essential.

While much of the recovery process is still ongoing, long-term recoveries post-disaster are quickly becoming the norm. Disasters are getting bigger. Climate change is occurring more rapidly—and disasters are intensifying. To ensure the Eastern Seaboard is better prepared for the next major storm, emergency managers and policy makers should utilize lessons learned to inform policy going forward.

CONCLUSION

The events presented over time illustrate many of the same challenges. The intention is not to show that emergency management is not learning lessons from each event—quite the contrary. The intent is to show that emergency management will continue to face similar challenges as they relate to vulnerable communities. While challenges like poverty, affordable housing, employment, market fluctuations, etc., will likely be ever present in all phases of disaster, how and at what rate these issues impact a specific community will always be unique. And, while it is certainly not the task of emergency managers to fix the socioeconomic and political problems or their causes, it is absolutely critical for emergency managers to have social intelligence about their community in order to know how the people emergency management interacts with will be able to be protected by its systems, and have access to aid and services post-disaster. Emergency managers need to know that preexisting conditions in families and communities will always be a determining factor in the success of preparedness, response, and recovery. Social, economic, and political conditions are inextricably linked to individual and community resiliency. Therefore, information on how these issues interact with the systems of a particular community must be part of emergency managers' tool kits. As all response is local and done within an all-hazards framework, it only makes sense that jurisdictional competence be a prerequisite for efficient and effective planning, response, and recovery.

Because the disaster aid given to people is intended to get them back on their feet, not to make them whole, it is critical to understand all the other factors individuals and communities contend with on a daily basis, as money to get a person back on his or her feet is going to be useless if he or she wasn't on his or her feet to begin with. Preexisting conditions are critical for an emergency manager to be aware of.

3

Interplay between Public Policy and Social Policy

In Chapter 2, we provide 10 case studies on major disaster events to show that the same issues present themselves after each disaster. That is, pre-existing conditions in a community drive what happens during and after a disaster. Unfortunately, in most instances, government and emergency management did not factor preexisting conditions into a tailored response and recovery. What does this mean for a community? It means that issues that exist prior to a disaster are exacerbated post-disaster and hinder resiliency for many communities. Not understanding or properly dealing with these preexisting conditions costs not only money but also lives. And over time, disasters will continue to hit communities and preexisting conditions will worsen. Ignoring these issues will continue a cycle of poor response and recovery efforts by emergency management.

This leads us to Chapter 3; in this chapter we will discuss the interplay between public policy and social policy and how these two sometimes competing structures combust during disaster. Disaster policy is social policy. Understanding this will help emergency managers understand how perceptions shape public policy. How government and society view people plays a significant role in electing our leaders, which results in laws and public policies. These laws and public policies are a reflection of the times. Once a disaster is added to the mix, the entire social fabric is exposed.

Historically, the fundamental question regarding disaster policy has been: How big of a role should government play prior to, during, and after a disaster? The public's view on this issue is constantly evolving because of the interplay between politics, public policy, and how the electorate views various communities.

> Citizens groups, members of Congress, and bureaucrats name particular groups of disaster victims as part of the American community and as deserving of help. Sometimes they argue for the community's obligation to victims of disaster, and at other times they protest that the bonds of sympathy are not strong enough to entangle the central government in what should be local and private matters. Whether or not the federal government intervenes is fought out in a kind of Hegelian dialectic for America between the ideals of a community's obligation to its members and the limited state. Over time, the democratic impulse prevails, and in disaster policy the impulse is a tendency to claim greater benefits for a greater number of people based on arguments about who is most deserving.*

Is disaster relief purely a function of the individual? Or should disaster relief be left to local, state, or nonprofit entities? Answers to these questions will vary depending on the political and philosophical leanings of the person answering. While this book respects the connection between the will of the people and the corresponding actions of the elected government, the foundation of disaster and social policy is rooted in how people perceive one another. Specifically, any group distributing resources always measures whether the group seeking help or relief is deserving of that help or whether there is a more deserving group. In sum, preconceived notions of the value of a group of people or perceptions of a group often drive how policies are developed and implemented. This is hardly an objective approach. This is where social intelligence can help dismantle poorly constructed perceptions in order to achieve a more robust end—a more resilient community.

Understanding that the political process is a reflection of the times can help emergency managers understand how a disaster impacts various populations. The goal of this chapter is not to change your thoughts about public policy or the political process, but instead to help you understand that sometimes public policy is shaped by perceptions (for better or for worse), and these perceptions shape how we prepare, respond to, and recover from disaster.

Think about this quote again: "We told you to leave. Why didn't you leave?" This quote is a result of a perception that many of the 100,000

---

* *Building the Disaster State: The Development of US Disaster Policy, 1789–1900* by Patrick S. Roberts Center for Public Administration and Policy -School of Public and International Affairs Paper prepared for presentation at the 2009 American Political Science Association Meeting in Toronto, CA, September 2–5, page 5.

people stranded during Katrina chose not to leave. Much of public dialogue was centered on irresponsibility. News pundits, government officials, and everyday people initially believed that 100,000 people made the irresponsible decision to stay behind and weather the storm. As discussed in Chapters 1 and 2, some preexisting conditions that prevented people from evacuating include

1. Many had nowhere to go.
2. Some did not have transportation.
3. Those that did have transportation did not have the financial resources to fill the gas tank or secure temporary shelter.
4. Many relied heavily on social networks (parish groups, extended family, friends) to make daily ends meet. Disconnecting from the social networks would devastate families.

These examples are just a small sampling of the complicated nature of why people make decisions about evacuating or sheltering in place. These same complications impact recovery. Isn't it about time emergency management takes these factors into account while developing emergency management systems? In order to develop more robust plans, we must understand that emergency management and emergency management policy are inextricably linked to public policy. And as stated before, public policy is a reflection of the electorate and elected officials.

As seen in the case studies, vulnerable populations are most severely impacted during and after disaster. The preexisting conditions—the interplay of social policy and vulnerable communities—are largely driven by perception. This chapter will look at the relationship between perceptions of vulnerable populations during and after disaster. You will see clear parallels and, in most instances, find that clichés and long-held beliefs about deserving and undeserving communities drive how government responds to disasters.

This chapter will illustrate how politics and perceptions shape social welfare policy. More importantly, this chapter does not lay blame on conservative, liberal, progressive, or any other political philosophy. Rather, the chapter illustrates how two centuries of politics and perception of the poor have shaped policies impacting the poor. Understanding the roots and evolution of public policy will help you understand how disaster relief policy developed. In sum, this chapter will provide a brief but detailed historical perspective on perceptions of the poor before and after disaster in addition to discussing how policy makers think about poor communities. Finally, the intent of the chapter is not to root out a specific political ideology. That said, in order to develop a social intelligence framework for your community or organization, it is important to understand how we got here and how we can integrate nuance into disaster response and recovery. A trillion dollars have already been spent—it's time to rethink how we spend the next trillion.

THE HISTORY OF DISASTER RELIEF AND SOCIAL WELFARE POLICY

Why are the connections between social welfare policy and disaster relief policy important to social intelligence? First, in order to provide a new perspective to emergency management, it is important to see the historical evolution of disaster policy. Research has shown that those who face the greatest impacts and have the least ability to recover from disaster in the United States are those who are in need or are on the verge of needing governmental assistance in daily life.[*]

Prior to the 19th century, the United States granted relief to various claimants via private bills.[†] By the mid-19th century, Congress had passed separate relief bills for such events as grasshopper plagues, Mississippi River floods, and the Civil and Indian Wars.[‡] Until this point, the U.S. Congress took action when it saw fit (when claimants were seen as blameless) and voted unanimously to provide relief to claimants. Michele Landis, author of *The Sympathetic State: Disaster Relief and the Origins of the American Welfare State*, argues that the origins of social welfare policy are rooted in the first appropriations of federal aid for disaster-type events. Landis argues that social welfare policy evolved from the federal government allocations for fires, floods, droughts, and other disasters. From the first private bills for relief of distress in the late 18th century, every successful claim enhanced the appeal of disaster relief as a precedent for those seeking federal funds.[§]

From the very beginning, Congress was very wary of expending federal dollars to aid disaster victims. Strong opposition always

---

[*] Meyer, Michelle A., Internal Environmental Displacement: Challenge to the United States, in *Welfare State Disaster and Sociolegal Studies*, ed. Susan Sterett.

[†] Dauber, Michele Landis, *The Sympathetic State: Disaster Relief and the Origins of the American Welfare State*, University of Chicago Press, Chicago, 2013.

[‡] Dauber, Michele Landis, *The Sympathetic State: Disaster Relief and the Origins of the American Welfare State*, University of Chicago Press, Chicago, 2013, location 522 of 9,055.

[§] Dauber, Michele Landis, *The Sympathetic State: Disaster Relief and the Origins of the American Welfare State*, University of Chicago Press, Chicago, 2013, location 874 of 9,055.

presented itself when the federal government attempted to provide relief post-disaster.

> Massachusetts Federalist Benjamin Goodhue complained about a request for relief after a 174 fire in Boston: "A fire happened lately in Boston, which destroyed ten or twenty thousand pounds worth of commodities that had paid duties. What kind of business would it be if all these persons were to come forward and demand compensation? ... Claims of this kind would never have any end."[*]

The quote above best exemplifies a historical attitude of Americans and their elected leaders toward disaster relief. Setting precedent was a consistent worry. Policy makers wondered aloud about paying one set of disaster victims and creating a permanent precedent that would expand the role of the federal government. Wrapped up in the philosophical debate surrounding the role of the federal government post-disaster, individual rights, and other governance issues was the idea of people seen as deserving or undeserving of help.

Does it make sense to determine fault during and after disaster? Who makes those judgments? For many years, Congress made those determinations. Each claim for disaster relief was handled individually. And during each debate, politicians debated the merits of the claim without any foundational knowledge of the population they were discussing.

Time after time, Congress made decisions about disaster relief based on its perceptions of a community. It debated whether disaster relief would result in dependency on government. Or whether one community deserved more help than another. This uneven and discriminatory policy was rooted in American thinking that government should not play a role in the life of the individual and, more importantly, that individuals going through difficult times have the wherewithal to lift themselves out of difficulty.

In Michele Landis's book *The Sympathetic State: Disaster Relief and the Origins of the American Welfare State*, she cites Harvard professor Alexander Keyssar:

> No one doubted that business panics occasionally threw some men and women out of work, but it was widely believed that these were transient episodes, affecting a small number of workers, who found new jobs in short order. If a worker was idled repeatedly or for a prolonged period of time, it was almost certainly his own choice or his own fault.[†]

---

[*] Sarat, Austin, *Catastrophe: Law, Politics, and the Humanitarian Impulse*, p. 65.
[†] Dauber, Michele Landis, *The Sympathetic State: Disaster Relief and the Origins of the American Welfare State*, University of Chicago Press, Chicago, 2013, location 1,168.

Landis goes on to cite Keyssar and write that this attitude began to change as depressions and other economic downturns impacted communities. Interestingly, Landis argues that the foundation for the New Deal was rooted in disaster relief policy.

Over time, the federal government increasingly involved itself in matters related to disaster relief; the consistent worry among lawmakers was to narrow relief so as not to encourage abuse or dependency. This philosophy and approach descends from the earlier Keyssar quote—a historical belief that individuals can lift themselves out of times of trouble.

The purpose of this chapter is not to argue for an influx of relief or for massive expansion of government. Instead, we argue that a construct based on beliefs that government relief should only go to deserving people guarantees failure during and after disaster. Allowing the allocation of disaster relief to follow this construct subjects disaster victims to the politics of the day. We believe there is a better way to deal with disaster policy—via social intelligence.

Later in this chapter, you will read about how disaster victims and current disaster relief policy are driven by perception. We have already argued that who we elect and the policies that result are directly correlated with how people view one another. More importantly, there are historical connections between disaster relief policy and general social welfare policy.

PROCESS OF DECLARING A DISASTER— STATE PROCESS AND THE STAFFORD ACT

In this section, we will provide a brief history of disaster policy. Understanding the process within the context of politics, perception, and preexisting conditions is important. Earlier in this chapter we argued that our public policy is a reflection of how we view people and communities.

The federal government provided no assistance to individuals until the Disaster Relief Act of 1974—the act created the Individual and Family Grant Program.* And until the passage of the Stafford Act in 1988, the

---

* Daniels, Richard R., *Vulnerability and Responsiveness in U.S. Disaster Policy: The Impact of the Stafford Act*, Conference Papers—Midwestern Political Science Association 1, 2009. Academic Search Complete, EBSCOhost (accessed September 21, 2013).

federal government only responded to natural disasters and occasional chemical and commercial disasters.*

So what has to happen for a state or municipality to receive disaster relief from the federal government? When disaster strikes, state and local governments assess the damage and determine the capability of the local jurisdiction and state to respond.† If the damage caused by the disaster greatly exceeds the capacity of the local government, the governor of the state declares a statewide disaster declaration. If the governor and the state emergency management agency decide the state cannot properly resource and respond to the disaster, the disaster relief request is made to the regional Federal Emergency Management Agency (FEMA) director. Next, the regional FEMA director conducts his or her assessment of the damage. Finally, the governor makes a written request to the president of the United States. The governor must indicate that every resource to assist the affected area has been exhausted or the disaster requires resources beyond the capacity of the state. After another review by FEMA, the president makes the final decision on whether to declare disaster.

POLITICS, PERCEPTION, DISASTER, AND SOCIAL INTELLIGENCE

The larger issue here is whether socially constructed views on deserving and undeserving populations are the best way to prepare, respond to, and recover from disaster. The answer is no.

The next section will provide in detail how perception and socially constructed views of the vulnerable impact disaster response and recovery policies and operations. Before you read the next section, think about this concept: When it comes to national security and understanding a nation's friends and foes, intelligence analysts focus on cold hard facts. That is, they study their target country's economics, population, geopolitical position, reputation, wealth, access to resources, and many other factors in order to gain some strategic understanding of the country. While politicians may hold certain beliefs and perceptions about the target country, intelligence analysts seek to provide a very matter-of-fact reading of the country based

---

* Daniels, Richard R., *Vulnerability and Responsiveness in U.S. Disaster Policy: The Impact of the Stafford Act*, Conference Papers—Midwestern Political Science Association 1, 2009. Academic Search Complete, EBSCOhost (accessed September 21, 2013).
† Ibid.

on various sets of information. Intelligence analysis also requires multiple viewpoints, and the target country is studied from a variety of angles. The goal: put our country's best foot forward via an objective approach. The politicians and policy makers are then free to add their subjectivity.

We don't use this process when writing domestic public policy. From the start, we use subjective frames to develop policies and programs. And this is certainly true in emergency management.

The next section will provide an exhaustive review of the connections between disaster policy, emergency management operations, socially vulnerable populations, and the social constructed views we place on top of disaster.

PERCEPTIONS OF POVERTY AND DISASTER

Perceptions are not facts; nevertheless, perceptions are the starting point for other attitudes and behaviors.* How people perceive their neighbors, other communities—disaster victims, the rich, the poor, and everything in between—impacts how governments develop policy and respond to the needs of their people. This idea is distinctively separate from politics. This chapter originated as part of our graduate work at the University of Chicago. The topic of perception, poverty, and disaster, and our conversations on this topic, often provoked strange looks at emergency management conferences, and our professors even questioned the relevance of perception in emergency management. We continued to push this conversation because we understood that public policy is rooted in how people think about other people. And if this was true outside of disasters, then it is true to how people and systems interact during and after disaster. The key difference: the disaster event magnifies these socially constructed beliefs and puts a spotlight on the systems that result from these beliefs. This does not mean one specific political ideology, party, or philosophy is superior and the other inferior. The purpose and importance of this chapter is that if we are to create a socially intelligent emergency management system, emergency managers must understand how preexisting conditions came to be. This means understanding the good, the bad, and the ugly. This

---

* Rahm, Dianne, and Christopher G. Reddick, US City Managers' Perceptions of Disaster Risks: Consequences for Urban Emergency Management, *Journal of Contingencies and Crisis Management* 19(3): 136–146, 2011. Academic Search Complete, EBSCOhost (accessed September 20, 2013).

includes acknowledging how historical trauma, institutionalized racism, and other structural and social problems are magnified during disaster.

Immediately after Hurricane Katrina we heard that the storm exposed the underbelly of New Orleans and America. The devastating storm only put a spotlight on everyday conditions in the city. If people didn't have the resources or transportation to evacuate at a moment's notice, they certainly didn't have those same resources in their daily life prior to disaster. Our point: the storm only brought attention to daily life in New Orleans with a disaster piled on.

Creating a socially intelligent emergency management system is not just about acknowledging social problems. Nor is it solely about responding to the structural issues surrounding poverty. Our social intelligence model is rooted in understanding the existing problems and making sense of how those problems were addressed. In most cases, it is simply acknowledging the social construction of the problem.

This chapter will shed light on the connections between preexisting perceptions surrounding vulnerable populations. These perceptions often shape elections and result in public policy. As a result of existing public policy, these perceptions help form opinions of communities and people during and after disaster.

In the introduction section of this book we wrote that the United States has spent $1 trillion on disaster response and recovery, and we are no more resilient after having spent that money than prior to the disaster. If we want better results from the next trillion, we must understand how we got here. How we got here is rooted in perception.

> We finally cleaned up public housing in New Orleans. We couldn't do it, but God did.
>
> —Rep. Richard Baker (R-La.), September 8, 2005*

PERCEPTIONS, REALITY, AND WHY WE CARE

Most of us remember the news coverage following Hurricane Katrina. After days of coverage where heroic images of the Coast Guard dominated were reports of all the ways evacuees had decided to take advantage of their situation and the federal aid system, in order to defraud the government and the

---

* Babington, C. (2005, September 10). Some GOP legislators hit jarring notes in addressing Katrina. *The Washington Post*, p. A04

American taxpayer for recovery money. As government reports about evacuee expenditures on jewelry, alcohol, and football tickets surfaced—and sensational media stories about purchases of guns, booze, and tattoos flooded the airwaves—access to recovery money became more difficult. Subsequently, those victimized by the storm and the inadequate disaster response became illustrations, through public discourse, of how America perceives the poor.

America saw on display uninformed assumptions and accusations based on commonly held beliefs about the poor. The public discourse was severe, not simple chatter at the water cooler or thoughts posted on personal blogs, but as statements vocalized by American leaders. Some suggested that conditions in the Astrodome were adequate as most people were poor anyway. Others displayed a complete lack of understanding regarding the ability of people to evacuate. And, many expressed a desire to punish people for their "bad choices." Some examples include

> What I'm hearing, which is sort of scary, is that they all want to stay in Texas. Everybody is so overwhelmed by the hospitality. And so many of the people in the arena here, you know, were underprivileged anyway so this [chuckle]—this is working very well for them.
>
> —Former First Lady Barbara Bush, on the hurricane evacuees at the Astrodome in Houston, September 5, 2005[*]

> I don't make judgments about why people chose not to leave, but, you know, there was a mandatory evacuation of New Orleans.
>
> —FEMA Director Michael Brown, arguing that the victims bear some responsibility[†]

> I mean, you have people who don't heed those warnings and then put people at risk as a result of not heeding those warnings. There may be a need to look at tougher penalties on those who decide to ride it out and understand that there are consequences to not leaving.
>
> —Sen. Rick Santorum (R-Pa.), September 6, 2005[‡]

---

[*] E&P Staff. (2005). Barbara Bush: things working out 'very well' for poor evacuees from New Orleans. *Editor & Publisher*, Retrieved April 16, 2009, from http://politicalhumor.about.com/gi/dynamic/offsite.htm?zi=1/XJ&sdn=politicalhumor&cdn=entertainment&tm=23&f=00&su=p504.1.336.ip_&tt=2&bt=0&bts=0&zu=http%3A//www.editorandpublisher.com/eandp/news/article_display.jsp%3Fvnu_content_id%3D1001054719

[†] *FEMA chief: victims bear some responsibility.* (2005). Retrieved April 16, 2009, from http://www.cnn.com/2005/WEATHER/09/01/katrina.fema.brown/

[‡] Kurtzman, D. (n.d). *Stupid quotes about Hurricane Katrina.*Retrieved April 15, 2009, from http://politicalhumor.about.com/od/currentevents/a/katrinaquotes.htm

How can you have the mess we have in New Orleans, and not have had deep investigations of . . . the failure of citizenship in the Ninth Ward, where 22,000 people were so uneducated and so unprepared, they literally couldn't get out of the way of a hurricane.

—Newt Gingrich[*]

The purpose of this chapter is to further discuss perceptions of the poor in America, reveal some of the realities of who is poor and why, and propose why it is critical information to consider in emergency management preparedness and operational planning.

Perceptions

Conversations about New Orleans post-Katrina paid acute attention to the Ninth Ward and its public housing residents and welfare recipients. In Chicago, during recovery operations, we often had to convince evacuees to pursue FEMA recovery money, for obtaining federal aid was foreign to most and embarrassing to many. One woman expressed her frustration at having to ask for assistance, as she had been the head of her household and "never asked for anything from anyone." She expressed the devastation she felt from both the storm itself and the fear of how she would be perceived when asking for help, as well as the desire to not be seen as a "welfare queen." She said, "I don't want the FEMA money, I want a job. A month ago I was a teacher, maybe poor, but a leader in my community. Today we are champagne-buying scam artists" (ABA conference participant, personal communication, February 10, 2006).

The fear of the welfare queen perception, post-Katrina, was not unfounded, given the public discourse at the time. However, in the disaster context, it is particularly poignant as this evacuee was eligible for disaster assistance for the exact same reasons as a wealthy New Orleanian, but expressed concern and frustration about pursuing the aid because of the long-standing public sentiment against the poor and utilization of public assistance.

The term *welfare queen* was created by former President Ronald Reagan while on the campaign trail. When delivering stump speeches in the 1970s, a major focus was on rolling back welfare. In order to illustrate

---

[*] Kromm, C. (2007). *Gulf watch: Gingrich at CPAC: New Orleans destroyed by lack of education, "citizenship."* Retrieved April 17, 2009, from http://www.southernstudies.org/2007/03/gingrich-at-cpac-new-orleans-destroyed.html

his reasoning for reducing social spending,* Reagan often told the story of a woman from the South Side of Chicago who drove a Cadillac and ripped off $150,000 from the government using 80 aliases, 30 addresses, a dozen social security cards, and 4 fictional dead husbands.† Given all the attention to this welfare queen, journalists attempted to find her in the hopes of interviewing her, but discovered she did not exist.‡

While the woman was not real, the image remained and helped lay the groundwork for subsequent poverty policy. So, even though the image was hyperbole, its impact was very real.§ According to Cassiman, the stories that we tell, the words we use, and the discourse of politics have shaped poverty policy as we know it in the United States. Utilizing words such as *welfare queen* conjures fear and dread, warranted or not. As welfare became a program associated with poor black mothers, "it became easier to attack," drawing on racist and misogynistic rhetoric.¶

The 8 years of the Reagan administration saw dramatic cuts to social spending, including Aid to Families with Dependent Children (AFDC), child care, unemployment insurance, legal aid, subsidized housing, and public and mental health services. As the cuts coincided with the recession, poverty and unemployment soared.** There was also a steep increase in the number of homeless people, which by the late 1980s had jumped to 600,000 on any given night, and 1.2 million over the course of a year. Many were veterans, children, and laid-off workers with no safety net. In a 1984 *Good Morning America* interview, Reagan attempted to defend himself against charges of callousness toward the poor and famously stated, "People who are sleeping on the grates . . . the homeless . . . are homeless, you might say, by choice."††

---

* Iceland, J. (2006). *Poverty in America: a handbook.* Berkeley and Los Angeles: University of California Press, Ltd., p. 4.
† Dreier, P. (2004). Reagan's legacy: homelessness in America. *National Housing Institute, 135.* Retrieved April 15, 2009, from http://www.nhi.org/online/issues/135/reagan.html, pg. 3.
‡ Ibid.
§ Cassiman, S.A. (2008). Resisting the neo-liberal poverty discourse: on constructing deadbeat dads and welfare queens. *Sociology Compass, 2*(5), 1690–1700. Retrieved April 15, 2009, from Wiley Interscience database. p. 1691.
¶ Cassiman, S.A. (2008). Resisting the neo-liberal poverty discourse: on constructing deadbeat dads and welfare queens. *Sociology Compass, 2*(5), 1690–1700. Retrieved April 15, 2009, from Wiley Interscience database. p. 1692.
** Iceland, J. (2006). *Poverty in America: a handbook.* Berkeley and Los Angeles: University of California Press, Ltd., p. 125.
†† Dreier, P. (2004). Reagan's legacy: homelessness in America. *National Housing Institute, 135.* Retrieved April 15, 2009, from http://www.nhi.org/online/issues/135/reagan.html, p. 4.

Reagan alludes in this comment to a popularly held belief that personal failings lead to poverty; if one worked hard enough, the system would allow him to pull himself up by his bootstraps, so to speak. This is a commonly held assumption, dating back to the beginnings of the nation. From the colonial period through the 19th century, the prevailing belief was that the roots of poverty lay chiefly not in structural economic causes, but in a person's individual misbehavior.* The poor were often seen as deserving or undeserving of public support.† The deserving tended to be children and the elderly. The undeserving were everyone else—the "idle able-bodied"—and were subject to indentured servitude, public whippings, or jail as a consequence to their time spent "idly or unprofitably."‡

The deserving/undeserving notion continued well into the 19th century, as the undeserving were considered "dependant, defective, and delinquent."§ Poorhouses opened as a method of punishing the poor. They were harsh, and the "inmates" were required to work as a form of punishment, moral training, education, and reform. Poorhouses fell out of favor in the beginning of the 20th century, as public officials began to realize that these types of institutions did little to reduce poverty and, in some cases, even exacerbated family instability.¶

In the beginning of the 20th century, some voices (e.g., authors, researchers, and Hull House social workers) speaking for the structural nature of poverty received attention. Harry Lurie, chairman of the American Association of Social Work Subcommittee on Federal Action, wrote in 1934 that "the dislocations and malfunctioning of the economic functions which produce unemployment and distress are not of an accidental character but are inherent in the nature of our economic organization."** While articulate, the structuralist argument did not win over most people, even at the height of the Great Depression. Well into the late 1930s, popular attitudes remained harsh, and polls showed the majority of people believed that "most poor people could get off relief if they tried hard enough." The poor were commonly referred to as "good-for-nothing-loafers" and "pampered

---

* Iceland, J. (2006). *Poverty in America: a handbook*. Berkeley and Los Angeles: University of California Press, Ltd., p. 11.
† Ibid.
‡ Ibid.
§ Iceland, J. (2006). *Poverty in America: a handbook*. Berkeley and Los Angeles: University of California Press, Ltd., p. 12.
¶ Ibid.
** Patterson, J. T. (2000). *America's struggle against poverty in the twentieth century.*Cambridge: Harvard University Press. pp. 43–44.

poverty rats."[*] Jokes of the time illustrate the lasting power of the lazy poor stereotype, such as: "There's a new cure for cancer, but they can't get any of it. It's sweat from a WPA worker." The initials WPA were often joked to mean "we pay for all" or "we putter around."[†]

Despite the attempts of the "war on poverty" in the 1960s, poverty persists—as do the perceptions of the poor.[‡] In 1975, one of the first large-scale studies of attributions of poverty was conducted. The researcher found that individualistic attributions were supported more strongly than other explanations, a finding that is "indicative of the national tendency to view poverty as a sign of personal and moral failure and the individualism characteristic of western cultures."[§] Other theories exist on the causes of poverty, such as structuralism, culturalism, and fatalism.[¶] However, as with attributions for poverty, individualistic explanations for wealth (e.g., drive, talent, hard work, risk taking) receive much greater support in the United States than structural acknowledgments. Thus, according to Bullock, "it is not simply the poor who are seen as deserving their economic status."[**]

In addition to the long history of the poor being perceived as lazy, uneducated, and morally askew, the poor today are often perceived to be nonwhite, and in most cases African American.[††] In a 1996 study investigating the accuracy of the media in portrayals of the poor between 1988 and 1992, it was found that the face of poverty was disproportionately black.[‡‡] While fewer than one-third of the poor are black, according to media portrayal, two out of every three poor people, or about 66%, are black. The study conducted by Martin Gilens also found that the deserving poor,

---

[*] Patterson, J. T. (2000). *America's struggle against poverty in the twentieth century.*Cambridge: Harvard University Press. p. 45.

[†] Ibid.

[‡] Appelbaum, L.D. (2001). The influence of perceived deservingness on policy decisions regarding aid to the poor [Electronic version]. *Political Psychology, 22*(3), 419–442. p. 419.

[§] Bullock, H. (2006). Justifying inequality: a social psychological analysis of beliefs about poverty and the poor. *National Poverty Center.* Retrieved April 15, 2009, from http://www.npc.edu/publications/working_papers/ p. 5.

[¶] Smith, K.B. & Stone, L.H. (1989). Rags, riches, and bootstraps: beliefs about the causes of wealth and poverty [Electronic version]. *The Sociological Quarterly, 30*(1), 93–107. p. 95.

[**] Bullock, H. (2006). Justifying inequality: a social psychological analysis of beliefs about poverty and the poor. *National Poverty Center.* Retrieved April 15, 2009, from http://www.npc.edu/publications/working_papers/ p. 7.

[††] Appelbaum, L.D. (2001). The influence of perceived deservingness on policy decisions regarding aid to the poor [Electronic version]. *Political Psychology, 22*(3), 419–442. p. 422.

[‡‡] Clawson, R.A. & Trice, R. (2000). Poverty as we know it: media portrayals of the poor [Electronic version]. *Public Opinion Quarterly, 64*, 53–64. p. 54.

especially the black deserving poor (e.g., working poor, elderly poor), were underrepresented in news magazines.*

In a 1990s survey about perceptions of the poverty population, respondents were asked, "What percent of all the poor people in this country would you say are black?" The median response was 50%. In a subsequent study a few years later, people were asked, "Of all the people who are poor in this country, are more of them black or white?" Fifty-five percent of the respondents thought more blacks than whites were poor, 24% thought more whites were poor, and the remaining considered the percentages to be equal. In interpreting his result, Gilens argues that this public view illustrates negative racial stereotypes, including the assumption that blacks are poor and lazy.†

Realities

Contrary to common perception and media portrayals, the poor are not disproportionately black, nor do they possess the characteristics attributed to the undeserving poor (e.g., lazy, bad decision making). In fact, the vast majority of poor Americans are white,‡ and those who obtain public cash assistance (Temporary Assistance for Needy Families (TANF)) are required to work and are only able to access this aid for a total of 5 years over their lifetime. The amount of TANF given to families depends on the state. In Illinois, in 2006, cash assistance for a family of three (only in a household with minor children) could receive a maximum of $396 (28.6% of the federal poverty line).§

In addition, according to Mark Rank, a social welfare professor at Washington University in St. Louis, one in three Americans will experience a full year of extreme poverty at some point in their adult life (extreme poverty is defined at 50% of the federal poverty line).¶ An esti-

---

* Clawson, R.A. & Trice, R. (2000). Poverty as we know it: media portrayals of the poor [Electronic version]. *Public Opinion Quarterly, 64,* 53–64. p. 54.
† Iceland, J. (2006). *Poverty in America: a handbook.* Berkeley and Los Angeles: University of California Press, Ltd., p. 39.
‡ U.S. Census Bureau.(2007). Income, Poverty, and Health Insurance Coverage in the United States. Retrieved April 14, 2009, from http://www.census.gov/prod/2008pubs/p60-235. pdf
§ National Center for Children in Poverty. *Illinois temporary assistance for needy families (TANF) cash assistance.* Retrieved May 27, 2009, from http://www.nccp.org/profiles/IL_profile_36.html
¶ Heartland Alliance for Human Needs and Human Rights. (2008). The 2008 report on Illinois poverty. Chicago, IL: Rynell, A. p. 7.

mated 58% of Americans between the ages of 20 and 75 will spend at least a year in poverty. Much like the disaster motto of "not if, but when," Rank also believes that, for most Americans, it appears it is also no longer if, but rather when, they will experience poverty. He says, "In short, poverty has become a routine and unfortunate part of the American life course."[*]

As of the most recent U.S. census data (2010–2011), 48.5 million Americans were living in poverty. This represents about 16% of the entire population. The national poverty numbers are determined by the federal poverty threshold.

It is also important to note that the federal poverty line is a minimal standard. There are many people considered poor or low-income who are not at or under the poverty threshold.

As discussed earlier in this chapter, poverty in America has generally been considered to be caused by individual failings (e.g., laziness, bad decision making). It seems we might hold so strongly to the premise of the American dream (i.e., "work hard and you shall receive") that we have been unable to accept that alternative theories could simultaneously exist. However, given the high rates of poverty, statistics that show most people will experience poverty at some point in their lifetime, and the reality that most people who are poor are working or are job seekers, it seems undeniable that other factors besides individual failings must contribute to poverty.

In 2000, when President George Bush announced his first cabinet, he stated, "People who work hard and make the right decisions in life can achieve anything they want in America."[†] What goes unrecognized by this prevalent yet simplistic view is the presence of (1) social failings and (2) structural failings in our economic system that can lead to poverty. Social failures such as sexism, racism, ageism, homophobia, xenophobia, etc., are issues American society continues to contend with. While the United States has certainly progressed in offering equal opportunities to all, companies and institutions continue to attempt to correct for social failures that permeate society with policy (e.g., affirmative action, minority hiring). For the purposes of this chapter, we will not focus on social issues as heavily as market failures.

---

[*] Pugh, T. (2007, February 26). Many Americans are Falling Deeper into the Depths of Poverty. *The Seattle Times*. Retrieved April 16, 2009, from http://seattletimes.nwsource.com/html/nationworld/2003589318_poverty26.html, p. 1.

[†] Henry, J.F. (2007). "Bad" decisions, poverty, and economic theory: the individualist and social perspectives in light of "the American myth." *The Forum for Social Economics, 36*, 17–27. doi: 10.1007/s12143-007-0005-z, p. 18.

The structural failure that contradicts the pervasive individual fail-ings theory for poverty is market failure. Adam Smith, often considered the father of modern economics, was the first major economic theorist to associ-ate poverty with the normal functioning of a capitalist economy. He said, "Wherever there is great property, there is great inequality. For one very rich man, there must be at least five hundred poor."* John F. Henry, economics professor, agrees. He writes, "Poverty is endemic to capitalism: given the normal workings of capitalism, poverty cannot be eliminated."† He sums up Smith's view on poverty as a societal outcome necessary to the foundations of capitalism and not the result of laziness or bad decisions, as "capitalism requires exploitation, and exploitation mandates a systemic division into rich and poor."‡ Simply put, laborers work for business owners who obtain a significantly larger percentage of the earnings from the labor. In addition, it is always better for the business owner to keep low-wage workers in order to maximize profit. This is not to say that economic growth is not beneficial in driving down poverty rates, but it certainly can be. It is good for business owners to be doing well and for the economy to thrive. This brings benefits to large segments of the population. However, even in times of great pros-perity, many Americans continue to experience poverty.

In short, we have chosen an economic system that works for many, but will always keep a percentage of people poor. As a result, there will always be people who live with increased vulnerabilities. For many, this is caused by the inability of the labor market to support all job seekers. Over the past 25 years, the American economy has experienced a substantial increase in the number of low-wage and part-time jobs.§ In fact, a 2000 study of U.S. full-time workers found that 25% of all American full-time workers could be classified as being employed in low-wage work, where low-wage is defined as earning less than 65% of the national median earn-ings for full-time jobs.¶

According to the 2007 poverty threshold chart (Table 3.1), if a family of four with two adults and two children had a full-time worker and a part-time worker in the household working (at a minimum wage of $6.55/ hour), the combined household earnings would keep the family at $591

---

* Henry, J.F. (2007). "Bad" decisions, poverty, and economic theory: the individualist and social perspectives in light of "the American myth." *The Forum for Social Economics, 36,* 17–27. doi: 10.1007/s12143-007-0005-z, p. 23.
† Ibid.
‡ Ibid.
§ Rank, M.R. (2004).*One nation, underprivileged.* New York: Oxford University Press. p. 54.
¶ Ibid.

below the poverty line of $21,027 gross income for the year. Studies show that a majority of the poor do have a family member in the labor market. A 1997 labor market study found that 37% of poor families were in full-time working families, and another 35% were in part-time working families.[*]

In addition to the low-wage work and the part-time work, there is also a deficiency in the number of jobs available to job seekers. Over 30 years ago, Milton Friedman, Nobel Prize-winning economist, helped develop the concept of the natural unemployment rate. Essentially, a degree of unemployment must be embedded in our economic system in order for free markets to function most effectively. Thus, a certain percentage of people must always be out of work.[†] So the idea that any person at any given time can just "pull himself or herself up by the bootstraps" is not true. There will be times when work, especially sustainable work or living wage work, is unavailable.

Why Is This Important for Emergency Management?

Why should emergency management concern itself with perceptions of the poor, the realities of being poor, and poverty data and trends? How is this consequential for emergency management?

This information is critical for emergency management because perceptions and misleading portrayals of the poor can have consequences. As Clawson and Trice argue in their media study of the poor, perceptions of poverty are important because they have an impact on public opinion. In turn, public opinion then has an impact on public policy. The study states, "If attitudes on poverty-related issues are driven by inaccurate and stereotypical portrayals of the poor, then the policies favored by the public (and political elites) may not adequately address the true problems of poverty."[‡] The study concludes that public opposition to welfare can in large part be attributed to misperceptions of the poor.

In the case of Katrina, miscalculations, gross misunderstandings, and negligence in New Orleans led to massive failures in evacuation planning and evacuation operations. After the storms, various politicians and

---

[*] Iceland, J. (2006). *Poverty in America: a handbook.* Berkeley and Los Angeles: University of California Press, Ltd., pp. 78–79.
[†] Blanchard, O. & Katz, L.F. (1997). What we know and do not know about the natural rate of unemployment. *The Journal of Economic Perspectives 11*(1), 51–72. Retrieved April 15, 2009, from Ebsco Host database.
[‡] Clawson, R.A. & Trice, R. (2000). Poverty as we know it: media portrayals of the poor [Electronic version]. *Public Opinion Quarterly, 64,* 53–64. p. 61.

Table 3.1 Poverty Thresholds for 2007 by Size of Family and Number of Related Children under 18 Years

Size of Family Unit	Weighted Average Thresholds	Related Children under 18 Years								
		None	1	2	3	4	5	6	7	8 or More
One person (unrelated individual)	10,590									
Under 65 years	10,787	10,787								
65 years and over	9,944	9,944								
Two people	13,540									
Householder under 65 years	13,954	13,884	14,291							
Householder 65 years and over	12,550	12,533	14,237							
Three people	16,530	16,218	16,689	16,705						
Four people	21,203	21,386	21,736	21,027	21,100					
Five people	25,080	25,791	26,166	25,364	24,744	24,366				
Six people	28,323	29,664	29,782	29,168	28,579	27,705	27,187			
Seven people	32,233	34,132	34,345	33,610	33,098	32,144	31,031	29,810		
Eight people	35,816	38,174	38,511	37,818	37,210	36,348	35,255	34,116	33,827	
Nine people or more	42,739	45,921	46,143	45,529	45,014	44,168	43,004	41,952	41,691	40,085

public officials commented on the lack of willingness of people to evacuate after the orders were given to do so, and cited this as a major contributor to the horrific aftermath. FEMA Director Mike Brown and Senator Rick Santorum suggested those who stayed in their homes bear responsibility for the results of not evacuating and should therefore be held accountable for their actions. Michael Chertoff, director of the Department of Homeland Security, felt similarly. He stated, "The critical thing was to get people out of [New Orleans] before the disaster. Some people chose not to obey that order. That was a mistake on their part."*

What we know now is that while some people did choose to ride out the storm, most people did not leave because they were not able to leave. When people talk about the Katrina evacuation, the complexities of the situation are often mentioned. However, for those that did not evacuate, the answer is quite simple—they did not have the money. In the words of those who did not evacuate:

> I didn't have gas in the car and I don't have the money. It's not like I'm trying to play tough. I just don't have the means to leave.
>
> —Tom Pendry, Biloxi, Mississippi

> Our family was already in a financially depressive situation before the hurricane. We had $300 between us. Mom had about $225 worth of savings. That was our emergency savings for anything. And that was a blessing. . . . It's hard to just get up and go when you don't have anything.
>
> —Jermaine Wise, New Orleans

> These people look at us and wonder why we stayed behind. Well, would they leave their grandparents and children behind? Look around and say, "See you later?" We had one vehicle. . . . They all couldn't fit in the truck. We had to decide on leaving family members—or staying. I'm living paycheck to paycheck.
>
> —Carmita Stephens

> Me and my wife, we were living paycheck to paycheck, like most everybody else in New Orleans. . . . I don't own a car. Me and my wife, we travel by bus, public transportation. The most money I ever have on me is $400. And that goes to pay the rent.

---

* Graham, S. (2006). Cities under siege: Katrina and the politics of Metropolitan America. *Social Science Research Council.* Retrieved April 18, 2009, from http://understandingkatrina.ssrc.org/Graham/, p. 1.

—Eric Dunbar, New Orleans

According to the *Washington Post*, Dunbar estimated his annual income to be about $20,000. Before the storm, when he and his wife estimated how much money they needed to flee the city, he realized that he could not come up with anywhere near the amount of money they would need for a rental car and airfare.*

Chertoff, Brown, and others immediately focused on the mistake of the residents who did not evacuate. It quickly became apparent, however, that the notable mistake was on the part of public officials who were clueless regarding the ability of many to self-evacuate and the underlying causes resulting in the inability to prepare and respond appropriately. In short, there was no good policy in place to address the needs of low-income residents. In fact, it seems that there was no policy at all. Some would probably argue that nobody cared to address this need, as most people were poor. Or as Wolf Blitzer described the evacuees, "Many of these people, almost all of them that we see are so poor and they are so black."† Others will argue that public officials just did not know. Either way, we now know that we need to be aware of how perceptions and misinformation can lead to poor policy or the lack of political will to address a specific issue.

In addition to being aware of perceptions and their effect on policy, it is equally important to marry that information with the realities of poverty in our communities. Specifically, understanding the increased vulnerability of a marked segment of the population, and where they are located or concentrated, is critical to emergency managers. As in every emergency, this information will affect the ability to respond and recover, protecting the greatest amount of life and property. It will also most certainly allow us to better know the true landscape for which we must prepare. As Craig Fugate (current FEMA director, previous director of emergency management for the State of Florida) stated while preparing for the 2006 storm season, "Poverty is the greatest impediment to preparation."‡

If we assume that it was not malicious intent or lack of caring that left people to die in New Orleans, we can conclude it was a lack of information

---

* Haygood, W. (2005, September 4). Living paycheck to paycheck made leaving impossible. *The Washington Post*, p. A33.
† Kurtzman, D. (n.d). *Stupid quotes about Hurricane Katrina.* Retrieved April 15, 2009, from http://politicalhumor.about.com/od/currentevents/a/katrinaquotes.htm
‡ Robinson, A. & McNeal, N.P. (2006, May 31). Poverty keeps many from preparing for storm Season. *The Miami Herald.* Retrieved August 30, 2006, from http://www.miami herald.com, p. 1.

and understanding of what it means to be poor, how many people are poor, the prevalence of poverty, and the impact these factors have on all phases of disaster. As Fothergill and Peek found in their research on poverty and disaster, while disasters have often been considered status levelers or "events that democratized the social structure," disasters do not arbitrarily distribute risk and vulnerability or eliminate preexisting socioeconomic conditions.* Many people face daily risks as a result of the social, political, and economic environment that will often result in increased vulnerability. The understanding that there is a strong correlation between the daily hazards one faces and the ability to prepare for, respond to, and recover from disasters has not gained widespread recognition.†

According to Fothergill and Peek's review and synthesis of the literature on poverty and disasters in U.S. populations over the last 20 years, socioeconomic status is a significant predictor of the impact of a disaster in both pre- and post-disaster stages. They studied risk perception, preparedness behavior, warning communication and response, physical impacts, psychological impacts, emergency response, recovery, and reconstruction. Their review concludes that the poor are more likely to perceive hazards as risky; less likely to respond to warnings; more likely to die, suffer injuries, and have proportionately higher material losses; more likely to have more psychological trauma; and more likely to face more obstacles during the phases of response, recovery, and reconstruction.‡ It is important to note that while it may seem contradictory for the poor to be more likely to perceive hazards as risky yet less likely to prepare, it makes sense, as preparedness requires economic and social resources that the poor often do not readily possess.§

The study goes into great depth highlighting both the invisibleness of the poor predisaster and the overwhelming task of response and recovery due to the lack of attention to preexisting vulnerabilities. According to an investigation of the effects of Hurricane Hugo on poor residents in South Carolina, it was found that because the rural poor were essentially invisible, living in unmarked homes on unmapped roads, or behind other homes, rescue workers did not know where people lived and that they needed assistance. Emergency response workers stated that until the

---

* Fothergill, A. & Peek, L. (2004). Poverty and disasters in the United States: a review of recent sociological findings [Electronic version]. *Natural Hazards, 32,* 89–110. p. 89
† Ibid.
‡ Fothergill, A. & Peek, L. (2004). Poverty and disasters in the United States: a review of recent sociological findings [Electronic version]. *Natural Hazards, 32,* 89–110. pp. 103–104
§ Fothergill, A. & Peek, L. (2004). Poverty and disasters in the United States: a review of recent sociological findings [Electronic version]. *Natural Hazards, 32,* 89–110. p. 104

storm, they had no idea of the extent of the poverty in their very own neighborhoods. A rescue worker stated, "There was poverty here all of the time, but after the storm I saw what I've been living around for 34 years."[*]

Their study also shows that the difficulty of recovery is exacerbated as poor households often lack access to resources and income needed to cope following disasters.[†] In addition, due to preexisting conditions, an event will often hit a poor family much harder. Fothergill and Peek explain that aid agencies (churches, community groups, government workers) are often not interested in dealing with the larger problems of poverty and want to limit assistance to disaster relief. For the poor, however, that is unrealistic. An outreach worker stated:

> I am still doing Hugo relief, but in all honesty, many of the things that we are doing are things that pre-existed. You cannot replace a roof on a wall that is rotted, and those are the types of things we ran into frequently because of the level of poverty.[‡]

The reality is that there are high and consistent rates of poverty. Working within an economic system that will always result in some level of unemployment and low-wage and intermittent workers, it is critical to take into account the information uncovered in this disaster research. We must consider the profound impact that poverty can have on all phases of disaster and discontinue the traditional blindness toward disaster vulnerability as a result of socioeconomic status. If we are ever to be truly resilient, we must realize disasters are not status levelers.

Finally, emergency management should concern itself with poverty data and trends, because not only the poor are affected by poverty. When poverty increases, especially at an increased time of unemployment (e.g., a local plant closure, outsourcing, government cuts in spending, recession, stock market variability), public services, especially emergency services, are often more heavily relied upon and utilized by more people. If we are to be strategic in not overwhelming emergency services and making sure that the most pressing needs are addressed, poverty rates and any other social data that indicate a potential decrease in resiliency and increase in vulnerability are crucial information. What they do, in short, is increase

---

[*] Fothergill, A. & Peek, L. (2004). Poverty and disasters in the United States: a review of recent sociological findings [Electronic version]. *Natural Hazards, 32,* 89–110. p. 97

[†] Fothergill, A. & Peek, L. (2004). Poverty and disasters in the United States: a review of recent sociological findings [Electronic version]. *Natural Hazards, 32,* 89–110. p. 98

[‡] Fothergill, A. & Peek, L. (2004). Poverty and disasters in the United States: a review of recent sociological findings [Electronic version]. *Natural Hazards, 32,* 89–110. p. 100

our situational awareness, pre- and post-event, a hallmark of good emergency management.

According to Rank, the country pays a heavy price for poverty. He sites a Children's Defense Fund report:

> The children who suffer poverty's effects are not its only victims. When children do not succeed as adults, all of society pays the price: businesses are able to find fewer good workers, consumers pay more for their goods, hospitals and health insurers spend more treating preventable illnesses, teachers spend more time on remediation and special education, private citizens feel less safe on the streets, governors hire more prison guards, mayors must pay to shelter homeless families, judges must hear more criminal, domestic, and other cases, taxpayers pay for problems that could have been prevented, fire and medical workers must respond to emergencies that never should have happened, and funeral directors must bury children who never should have died.[*]

In short, if communities are to best respond to an increase in the need for fire, police, emergency rooms, etc., there needs to be as much information as possible about the need.

Leaders in emergency management often place a strong focus on forward thinking for the most comprehensive emergency management planning. Imagine if someone had been forward thinking enough just to consider the increased vulnerability posed by poverty and coupled that with socioeconomic data. Even a small amount of social data would have shown that in 2005, 12.4% of Americans were living in poverty, a not insignificant number. In the case of New Orleans, however, it was even greater, with 27.9% of people living at or below the poverty line, almost one in three people, and many others living just above or close to the poverty line, or having an income low enough to be eligible for public assistance.[†] Imagine if we were able to go beyond simply knowing the socioeconomic landscape and had a mechanism within emergency management to deploy resources to the most vulnerable communities in a mitigation capacity.

Knowing this type of information gives emergency management professionals a sounder grounding in poverty, and thereby arms the key guardians of life and property with the important, on-the-ground information. It is the same type of information that community policing utilizes, the same type of information intelligence analysts depend on, and

---

[*] Rank, M.R. (2004). *One nation, underprivileged*. New York: Oxford University Press. p. 110

[†] The Urban Institute. (2005). *Katrina: demographics of a disaster*. Washington, D.C. Retrieved April 20, 2009, from http://www.urban.org/UploadedPDF/900835_katrina_factsheet.pdf

it is certainly critical to better understanding of "all hazards" in its truest form. Authors of a well-known 1994 poverty and disaster study found: "Indeed, disasters are the products of the social, political, and economic environment, as well as the natural events that cause them."* In sum, the question is not why emergency management should care about historical perceptions of the poor, current poverty data, and their implications; the question is how we can ignore information so central to the fundamental mission of emergency management.

---

* Fothergill, A. & Peek, L. (2004). Poverty and disasters in the United States: a review of recent sociological findings [Electronic version]. *Natural Hazards, 32,* 89–110. p. 89

Section II

Social Intelligence Framework and Intelligence Methodology and Emergency Management

4

Intelligence Methodology and Emergency Management

Before beginning this section on intelligence, we want you to know one thing: we are not advocating for the collection of intelligence on domestic populations. Simply, this is not a method to utilize or justify spying. Our social intelligence (SI) model is based on emergency management aggregating already available information (e.g., unemployment rates, foreclosure rates, homeless numbers) and using it in ways to better prepare for, mitigate, respond to, and recover from disasters. The purpose is to gather quantitative data about a community for emergency managers so that they can engage in qualitative assessments regarding the interplay of the data with emergency management systems.

The national security apparatus uses a similar approach to add value to national security policy and U.S. international interests. That is, these agencies provide policy makers with historical and real-time information on geopolitics, economics, and other issues that impact the region they are interested in. This same approach should be used to better understand and serve domestic populations in preparing for responding to disasters. SI is simply leveraging intelligence methodology in peaceful ways such that emergency managers and policy makers are provided with robust information on the populations they serve. This includes the ability to learn from disaster events and daily life.

Intelligence agencies work to understand people, their environments, and the systems they interact with by connecting informational dots. Making these connections creates an intelligent and living system that leverages information, human expertise, and technology to provide situational awareness. This situational awareness provides policy makers with enough information to make educated and informed critical decisions. This information can also help policy makers ask different or better questions.

We now know that perceptions of people and communities often contribute to who we elect, how policy is formed, and how it is implemented. If we recognize this, we must also recognize that emergency management policy is also a reflection of perception and policies based on those perceptions. What emergency management needs is a methodology to study preexisting conditions in a community prior to disaster, which strips away all of that. Intelligence analysts present their cases from multiple angles and acknowledge historical issues along with the politics and geopolitics of the region. They do this to paint a complete picture. How does this benefit an emergency manager? Understanding what is happening in a community on the front end will help emergency managers develop more informed emergency management systems. To implement social intelligence, emergency managers must have a foundational understanding of the types of social intelligence available, where to gather it, how to make sense of it, and how to apply that information to develop a socially intelligent emergency management system.

This section of the book will provide you with a comprehensive understanding of U.S. national security and intelligence apparatus. This background information can be leveraged with our social intelligence model to help you create a socially intelligent emergency management system for your community.

INTELLIGENCE METHODOLOGY AND EMERGENCY MANAGEMENT

There is a role for intelligence methodology in emergency management. Social intelligence relies on various qualitative and quantitative data points to better understand a community prior to disaster. This knowledge is what we call jurisdictional competence. In Chapter 3, we argue that perceptions play an important role in driving public policy and programs. The interplay between perception and the politics that result from perceptions is what has led to inefficient and unintelligent disaster responses. Clearly, there is an abundant amount of information from lessons learned (during previous disasters), from historical and academic works, and from bringing data that currently exist in communities to bear. In the aggregate, this amounts to intelligence. Our point is very simple: just as militaries and intelligence agencies seek to understand the regions, countries, and people prior to taking on a massive operation, emergency managers must do a better job of understanding the communities they serve. Intelligence operations are a known and time-tested way of understanding communities. Understanding the history, culture, food, and political context of a specific country can be applied domestically. In sum, there are some fundamental truths—outside of politics—that if understood, help policy makers get a foundational understanding of a community. In this section, we will provide some historical context in order to lay the foundation for the social intelligence framework.

While social intelligence is a peaceful and domestic application of intelligence, know that intelligence has been used for millennia by countries to gain a better understanding of their neighbors and rivals. For centuries, people have searched for ways to gain a tactical advantage over their enemy. Most societies accomplished this by gathering information on their target, studying the information, and then using the information toward the best possible advantage over their enemy. The United States has created an extensive network of intelligence agencies designed to protect the nation from both internal and external threats. In a speech by Dennis C. Blair, the former U.S. Director of National Intelligence, he presents four fundamental goals of the intelligence community: "The first is to enable wise national security policies. . . . Our second goal is to support effective national security action. . . . Our third goal is to deliver balanced and improving capabilities so that the future intelligence community can be even more effective than today's. . . . And our fourth strategic goal is to function as an integrated team."* As the field of emergency management looks toward the future, it is our obligation to explore new ways to mitigate the effects of disasters against all life and property. We believe that by applying the core methods employed by the intelligence community, emergency management can further its capabilities and better identify the needs of the communities they serve. It is our intent to demonstrate how the field of emergency management can utilize intelligence methodology and situational awareness to mitigate the effects of disasters by constructing a robust planning system for complex events.

---

* Blair, D.C., Road Rules for the Intelligence Community, *Vital Speeches of the Day* 76(4): 176–183, 2010.

HISTORY OF INTELLIGENCE

The art of intelligence can be traced back thousands of years to the beginning of written history. During his reign as king of Babylonia, Hammurabi established an intelligence agency dedicated to gathering information on neighboring tribes and distant enemies.[*] In addition to Hammurabi, there are examples of intelligence operations documented throughout Egyptian history, Homer's *Odyssey* and the Bible.[†] The earliest known record of espionage is found in an ancient Egyptian text during the battle of Kadesh (1274 BC), when the pharaoh Rameses was deceived by two spies and unknowingly sent part of his army into an ambush. The pharaoh was fortunate though because he captured two more spies, and through interrogation, the plot was exposed and the pharaoh was able to avert disaster.[‡] In the epic Greek poem, *The Odyssey*, Homer writes, "But now change your theme and sing to us of the stratagem of the Wooden Horse, which Epeius built with Athene's help, and which the good Odysseus contrived to get taken one day into the citadel of Troy as an ambush, manned by the warriors who then sacked the town."[§] The Bible tells us in the book of Numbers, while leading the Israelites out of Egypt, Moses sent one leader from each ancestral tribe to scout the land of Canaan before entering.[¶] Moses was interested in gathering information about the people and landscape of the land of Canaan.

Intelligence was primarily born out of a desire by military commanders to identify weaknesses in enemy forces in order to defeat their enemies on the field of battle. In the *Art of War*, Sun Tzu writes, "Now the reason the enlightened prince and the wise general conquer the enemy wherever they move and their achievements surpass those of ordinary men is foreknowledge."[**] With the ability to know an enemy force's size, weaponry, and positions, a commander of an opposing force is more able to prepare his soldiers for battle and has a greater chance of victory. However, the use of intelligence was still not prevalent because commanders did not view it as necessary for victory on the battlefield. It was not until the end of the 18th century, when the industrial revolution brought about the

---

[*] Mulrine, A., and Bentrup, N., The Power of Secrets, *United States News & World Report* 134(3): 48, 2003.
[†] Columbia Electronic Encyclopedia, Office of Strategic Services, 2009.
[‡] Crowdy, T., *The Enemy Within: A History of Spies, Spymasters and Espionage*, Osprey Publishing, 2006.
[§] Homer, *The Odyssey*, trans. E.V. Rieu and D.C. Rieu, Penguin Group, London, 2003.
[¶] Bible, Numbers 13.
[**] Tzu, S., *The Art of War*, trans. S.B. Griffith, Oxford University Press, New York, 1963.

modern age of intelligence, that governments began to realize the significant role intelligence could play during military campaigns. In the article, "An Historical Theory of Intelligence," Kahn writes, "Industrial output was nationalized. Suddenly, factors that had never counted in war became significant. It mattered little to a medieval king how much coal and iron his enemy could produce; such knowledge was vital to a modern head of state. Railroads made possible the rapid mobilization, concentration, and supply of large bodies of troops. These deployments called for war plans far more detailed than any ever envisioned.... At last, intelligence had targets that gave it a chance to play a major role in war."* As we examine the history of intelligence, we observe how it developed as a weapon in support of militaries against neighboring and foreign adversaries. As travel became more prevalent and the idea of globalization advanced, intelligence began to play a more important role due to the magnitude of information available as societies continued to grow along with advancements in technology. By the end of the 18th century, the foundation of American intelligence had been laid and was well on its way to becoming the modern intelligence service that has been achieved today.

HISTORY OF U.S. INTELLIGENCE

In the United States, the founding fathers understood the importance of intelligence. After an apparent ambush during the French and Indian War, George Washington wrote, "There is nothing more necessary than good intelligence to frustrate a designing enemy, & nothing that requires greater pains to obtain."† As colonists in America became more and more frustrated by colonial rule, they organized into an underground anti-England society and adopted the name Sons of Liberty. Credited with being America's first spy ring, the Sons of Liberty were established in every major town across the 13 colonies. They first used pamphlets and newspapers to spread propaganda throughout the British colonies to fuel discussions of revolution. As the American Revolution began, the Sons of Liberty continued their work by scouting enemy forces, passing along intelligence, and infiltrating pro-British organizations.‡

---

* Kahn, D., An Historical Theory of Intelligence, *Intelligence and National Security* 16(3): 81, 2001.
† Allen, T.B., George Washington, Spymaster: How the Americans Outspied the British and Won the Revolutionary War, *National Geographic*, 2004, pp. 16–17.
‡ Tzu, S., *The Art of War*, trans. S.B. Griffith, Oxford University Press, New York, 1963.

After the war, intelligence operations were not considered vital during peacetime and were only used sporadically until the beginning of the Civil War. During the Civil War there was no central agency that controlled intelligence operations for either side. However, both the Confederate and Union armies participated in intelligence activities on a regular basis. Ethier writes, "In the absence of organized intelligence-gathering bodies, information-hungry commanders counted on tried and true sources such as prisoners, deserters, newspapers...and the civilian network."* Both sides of the conflict also tested the use of hot air balloons to scout troop movements and positions. Although ultimately deemed ineffective, this is considered to be the forerunner of aerial surveillance. Yet again, at the end of the war, intelligence operations were disbanded and much of what was learned throughout the war dismissed. The first formal military intelligence unit of the United States was the Office of Naval Intelligence in 1882, shortly followed by the Army's Military Intelligence Division in 1885. Officers from both services were assigned to foreign embassies throughout the world and primarily tasked with the collection of open-source information regarding their host country.[†]

During World War I the State Department, along with other intelligence agencies, was responsible for intelligence operations. However, by the time the United States entered the war and adequate funding was provided to the agencies, they lagged too far behind to have a big enough impact on the war. Once again, when the war ended, funding was taken away from intelligence programs and many of the offices were disbanded. In 1941, the first peacetime civilian-operated intelligence agency, Office of the Coordinator of Information, was established.[‡] However, several missteps in the office led to the bombing of Pearl Harbor on December 7, 1941. Once the United States entered World War II, the Office of the Coordinator of Information was absorbed into the Office of Strategic Service in 1942, which controlled intelligence operations for the United States during the war. Once again, at the end of World War II, the need for a peacetime intelligence agency was seen as no great cause for concern and the Office of Strategic Services was disbanded in 1945.[§] However, it was not long until President Harry Truman realized the need for intelligence during early

---

* Ethier, E., Intelligence: The Secret War within the War, *Civil War Times* 46(2): 15–17, 2007.
[†] Central Intelligence Agency, *History of American Intelligence*, April 15, 2007. Retrieved September 12, 2010, from www.cia.gov/kids-page/6-12th-grade/operation-history/history-of-american-intelligence.html.
[‡] Tzu, S., *The Art of War*, trans. S.B. Griffith, Oxford University Press, New York, 1963.
[§] Columbia Electronic Encyclopedia, Office of Strategic Services, 2009.

phases of the Cold War. In 1946 President Truman established the Central Intelligence Group. However, almost 2 years later the Central Intelligence Group was disbanded under the provisions of the National Security Act of 1947, and the National Security Council and the Central Intelligence Agency were created.* The era of modern U.S. intelligence had begun.

In this section, we provided a global and historical snapshot of intelligence and how it's been used. We also provided a brief historical timeline for U.S. intelligence operations. Why? The purpose is to lay out a philosophy and a framework for how emergency management can use principles of intelligence operations to develop a social intelligence system for their communities.

The goal of intelligence (as it is used in military and foreign policy) is to better understand a friend or foe. More information is better than having less information. And in the world of foreign policy or military affairs, more information can be the difference between life and death. The same is true in emergency management, except the goal of collecting information (social intelligence) is not to spy or surveil populations. The goal is to better understand communities prior to disaster through a process that is somewhat divorced from the social construction of how people view the communities they live in.

The goal with any intelligence operation is to collect information, analyze information, understand the information in the aggregate, and then act on information. *Actionable intelligence* is a term applied to information that policy makers can act on.

COMPREHENSIVE STUDY—GEOPOLITICAL AND SOCIOECONOMIC LANDSCAPE

So what qualifies as intelligence? What type of information helps policy makers make decisions? Examples include

- Population
- Regional actors—allies/adversaries
- Economic activity and drivers
- Religious factors, e.g., sectarian violence, religious divisions, etc.
- Access to rivers, ports, oceans
- Assessment of the internal supply chain

---

* Tzu, S., *The Art of War*, trans. S.B. Griffith, Oxford University Press, New York, 1963.

- Foreign exports
- Historical context related to current intelligence operation

The list can go on for pages. The point is to show how a variety of factors shape a country's perception and actions toward another. These factors must be weighed against one another. Prior to making a major decision, a policy maker must understand the intended and unintended consequences of his or her decision. Taking an action against one nation could impact the supply chain of another, setting off a chain of positive or negative world events.

Understanding the connections between one decision and the impacts of that decision on other factors is what makes intelligence operations intelligent. It is true that the value of a decision (both good and bad) is correlated to the value or validity of the intelligence. There is no doubt about that. This is why intelligence agencies seek to collect as much information as they can, some directly related to their study target, and some purely for contextual reasons. Why? Connecting the dots requires understanding the bigger picture—understanding the full landscape.

This same framework can be applied to emergency management. If emergency managers better understand the connections between a lack of affordable housing predisaster, they are better prepared to develop emergency shelter plans immediately after disaster and begin thinking about long-term recovery and housing as life safety operations come to an end. Similarly, a historical distrust of government led to a botched mandatory evacuation order in New Orleans during Hurricane Katrina. Another factor not considered: lack of access to financial instruments—credit cards, checking accounts, etc. A lack of access to these instruments and a historical mistrust of government led to over 100,000 people in New Orleans becoming stranded in the aftermath of Katrina. These preexisting conditions are forms of intelligence—social intelligence. Had emergency managers been acutely aware of these issues, they could have adjusted their emergency management systems prior to disaster instead of reacting to these issues during a catastrophic storm.

During the *Deepwater Horizon* oil spill of 2010, social intelligence could have helped emergency managers align disaster response operations and relief aid in a more coordinated fashion. In this case, the ocean was the company town. Every facet of life in the Gulf Coast was connected to water. Gaming, tourism, fishing (commercial and sport), fish processing, oil production and refinery operations, and all the supporting industries were linked to the health and vitality of the ocean. If the ocean were removed from the equation, the regional economy would be immediately impacted. And as each day passed, the regional economy suffered. The cascade of impacts could have been predicted.

Collecting more information will not prevent preexisting conditions from impacting disaster operations. In fact, in most instances, emergency managers and local governments will not be able to address or solve pre-existing conditions prior to disaster. That said, they can begin to understand that some issues persist because of a socially constructed view on a social problem. Some issues persist because of a structural failure in our economy. And some issues persist because of a lack of resources.

INTELLIGENCE IN ACTION IN EMERGENCY MANAGEMENT

Napoleon Bonaparte once said, "If I always appear prepared, it is because before entering on an undertaking, I have meditated for long and have foreseen what may occur. It is not genius which reveals to me suddenly and secretly what I should do in circumstances unexpected by others, it is thought and meditation."

Throughout the course of this book, we've discussed why collecting and understanding information about communities is a much better way to prepare for, respond to, and recover from disasters. The question is how? What information should emergency managers collect? Which information should they use? How should information be compiled and analyzed? These are great questions. The goal is to use various data points to develop actionable intelligence. We define *actionable intelligence* as the aggregation of various forms of data and information used domestically to inform social intelligence operations.

But before we discuss the implementation of social intelligence, the cycle of intelligence gathering must be discussed. The next section will focus on the cycle of intelligence as it currently exists in a national security context. Then we will discuss how this cycle can be used to implement social intelligence in a community.

---

* Builder, C.H., S.C. Bankes, and R. Nordin, *Command Concepts: A Theory Derived from the Practice of Command and Control*, RAND, Santa Monica, CA, 1999.

INTELLIGENCE CYCLE

The intelligence cycle (Figure 4.1) as defined by the Federal Bureau of Investigation (FBI) is "the process of developing unrefined data into polished intelligence for the use of policy makers."[*] The intelligence cycle is cyclical in nature and consists of six steps identified and defined below. However, in some instances the discovery of new information may require one to go back to an earlier step before he or she can move forward in the intelligence cycle.

Here, we will define each step in the intelligence cycle to give the reader an idea of how the intelligence process works.

- **Planning:** The management of the intelligence effort. This includes anything from identifying the need for the intelligence collected to delivering the actionable intelligence to the consumer. Many times this is both the beginning and end of the intelligence cycle because actionable intelligence often produces new requirements.
- **Collection:** The collection of raw data based on the requirements of intelligence. This is accomplished by acquiring information from a variety of sources.
- **Processing:** The process of translating the immense amount of data collected into a usable form for analysis. This is accomplished by decryption, language translations, and data reductions. The raw data are then processed into databases where they can then be used for analysis.
- **Analysis:** The raw data are assembled, analyzed, and then converted into actionable intelligence for use by policy and decision makers. The finished intelligence reports provide assessments of the information and its implications to the requesting authority.
- **Dissemination:** The final step is the distribution of actionable intelligence to the consumer. The actionable intelligence delivered is used in support of policy and decision making. Often, dissemination of actionable intelligence leads to further requirements by the consumer and, as a result, the continuation of the intelligence cycle.

---

[*] Federal Bureau of Investigation, *Intelligence Cycle*, n.d. Retrieved November 30, 2010, from http://www.fbi.gov/about-us/intelligence/intelligence-cycle.

Figure 4.1   Illustration of the intelligence cycle.

The intelligence cycle depicted above represents how the intelligence community operates to formulate actionable intelligence. Now you have a foundational understanding of the connection between intelligence and emergency management. In Section III of this book, we detail how an intelligence framework can be applied to emergency management to implement social intelligence.

Section III
Program and Policy Prescriptions

5

Implementing Social Intelligence

In Sections I and II we provided an overview of why emergency management must evolve. This final section of the book will provide a blueprint for implementing social intelligence in your community or organization. In addition, this section will continue to drill down into the rationale for implementing social intelligence in all communities.

We also want to emphasize that social intelligence is not a system for large cities. Implementing social intelligence will not require multi-million-dollar IT systems or software tools. Implementing social intelligence simply requires emergency managers to do what they do best: convene conversations, gather information, and develop plans and systems based on more localized, relevant demographic information.

Implementing social intelligence is a matter of understanding what information to look for, where to obtain the information, how to make sense of it, and then taking action using that information. However, in most instances the issues impacting disaster vulnerability are linked to historical and structural issues. Emergency management is not the vehicle to solve these problems, but it is the vehicle to understand these issues as they impact people before disaster strikes and as they are magnified once disaster hits a community. There is no escaping this reality. Chapter 2 illustrated that regardless of the type of disaster, the same issues keep impacting emergency management systems post-disaster.

> The traditional method of planning for emergencies including the stages of preparedness, response, recovery, and mitigation is viewed by some as outdated in today's highly vulnerable environment. For example, this life cycle approach is deterministic and the phases often overlap and blend together. In comprehensive vulnerability management, there is an

identification of vulnerabilities for communities and efforts are used to reduce these risks. This approach is more proactive, rather than reactive, as seen through the different phases of emergency management because it considers the most likely risks for a community.[*]

We estimate that the total cost of disasters to the American taxpayer and economy is $1 trillion. And we are repeating again that we should be smart about how we spend the next trillion. Hurricane Katrina alone cost the U.S. taxpayer $250 billion. Today, there are still millions of taxpayer dollars flowing into New Orleans to fund various recovery efforts. These resources, while critical, are doing very little to strengthen social systems and physical infrastructure in a way to save disaster dollars in the future. In 2010, the U.S. taxpayers shelled out over $300 million to aid in recovery efforts for flooding in Cook County, Illinois. Those dollars are going toward grants and low-interest loans to help individuals repair their homes. If local officials had known predisaster that many of the affected individuals did not have adequate flood insurance, could not afford debris removal or sump pumps, the taxpayer would be on the hook for less. Emergency management can do better.

Emergency management is not in the business of protecting the taxpayer's bottom line. The field does have a role in addressing preexisting vulnerabilities so that smaller disasters do not become catastrophic events. A rainstorm should not cost the taxpayer $300 million. After spending $250 billion in New Orleans, the levee system should be able to withstand a category 4 hurricane. The money has already been spent, and in many cases, we are no better off now than prior to disaster. That said, it isn't for a lack of trying—emergency management's problems stem from a lack of understanding.

We believe that emergency managers need to increase their knowledge about the communities they serve. The goal is to ensure that emergency managers are jurisdictionally competent. That is, emergency managers have access to various social intelligence data points to better inform their emergency management systems prior to disaster.

In Section II you learned about the intelligence field. You learned that in the national security world intelligence is about gaining a tactical advantage over the enemy. You may remember reading this in Section

---

[*] Rahm, Dianne, and Christopher G. Reddick, US City Managers' Perceptions of Disaster Risks: Consequences for Urban Emergency Management, *Journal of Contingencies and Crisis Management* 19(3): 136–146, 2011. Academic Search Complete, EBSCOhost (accessed September 21, 2013).

II, "Industrial output was nationalized. Suddenly, factors that had never counted in war became significant. It mattered little to a medieval king how much coal and iron his enemy could produce; such knowledge was vital to a modern head of state. Railroads made possible the rapid mobilization, concentration, and supply of large bodies of troops. These deployments called for war plans far more detailed than any ever envisioned . . . , At last, intelligence had targets that gave it a chance to play a major role in war."[*] In emergency management, and when discussing social intelligence, we are not talking about enemy states or actors. We are not talking about spying on Americans or our neighbors. *We are talking about using the same methodologies used in our national security apparatus to better understand the people and communities we serve and are obliged to protect.*

Unfortunately, policy makers appear hesitant to embark on long-term strategies or initiatives to systematically address disaster preparedness, mitigation, response, and recovery in poor communities.[†] Donner and Rodriguez add that the massive devastation caused by Katrina was not enough to transform disaster policy—the authors go on to state that unless there is a reshaping of political discourse on vulnerability and disaster, the field will continue to fail people as vulnerabilities increase. Scholars have been writing for decades about social vulnerability, poverty, and inequality. There are hundreds of books written on vulnerability and poverty. But we are still having the same politically charged discussions about vulnerability. We believe that in order to have the right conversation, administrators, elected officials, and emergency managers must know more about people. They must understand that social vulnerability is about more than just income, poverty, or jobs. Vulnerability can be as simple as not having access to credit instruments or culturally competent disaster communications. It also means that vulnerable populations can look like a middle-class family—a family with stable income, but little savings. Social vulnerability is not just about poverty. Social vulnerability is a spectrum that includes the very poor to the well resourced. Understanding how this spectrum impacts emergency management operations and policies is essential to implementing social intelligence.

If we are going to have a proper discussion about the future of emergency management and disaster policy, we have to start by understanding how things work from a systems perspective. Many public policy debates fail

---

[*] Kahn, David, An Historical Theory of Intelligence, *Intelligence and National Security* 16(3): 81, 20.

[†] Donner, W. and H. Rodriguez. "Population Composition, Migration, and Vulnerability." *Social Forces.* 87: 1089–1114.

because people fail to see the larger system. Public policy solutions should focus on how one change within a component causes ripple effects throughout a system. A chain is only as strong as its weakest link—similarly, public policy debates should focus on entire systems rather than individual parts.

Information is critical for any emergency manager. Too often the information required for robust emergency management often goes overlooked. The response and recovery failures of Hurricane Katrina illustrate that point. *Even if a policy maker or emergency manager thinks he or she knows his or her community well, there are always tools to be used, and information to be gathered, that can help community understanding and policy effectiveness.*

To gain a greater understanding of a community, jurisdictional competence must be developed before disaster strikes. To achieve greater understanding, we must begin to ask more questions. Then we must discern what the right questions are. We are reincluding our definition of *jurisdictional competence* from Chapter 1 for reference:

> Jurisdictional competency is achieved by developing socially intelligent emergency management systems. This means emergency managers, policy makers, and elected officials understand how the preexisting conditions in a community drive what happens during and after disaster. More importantly, once an emergency management system is socially intelligent and subsequently jurisdictionally competent, it understands how structural issues (social capital, poverty, government revenues, schools, demographics, etc.) are all connected. In understanding these connections, emergency managers can tailor emergency response and recovery resources in a manner that helps all people.

But to do this, we must first leverage information across all sectors of a community. Emergency managers bring together their resource partners in the emergency operations center (EOC). It is in the EOC that partners in response and recovery work together to identify assets and supply chain mechanisms to allocate resources. Many of these assets are interconnected. They are dependent—either upstream or downstream—on an incident spectrum. For example, a transportation system is reliant on effective utility access. Social service providers may require public safety cooperation. Most governments maintain a variety of information centers. Nearly every department maintains mechanisms to manage information—these databases contain information related to budgets, personnel, grants, infrastructure, and service. Discussions in the EOC foster a cross-functional dialogue, but preexisting conditions that organizations face prior to disaster impact critical assets after disaster. Understanding those preexisting conditions is critical and information driven.

Disasters occur when they impact human populations by way of impacting social systems and infrastructure. How people react to disaster is directly related to their jurisdictional capital. To understand the full value of jurisdictional capital, emergency managers must have a framework to map infrastructure with social systems.

The foundation of social intelligence is to identify, aggregate, and synthesize the relationships and interdependencies of baseline data points (BDPs). To understand the interconnections between the public, private, and nonprofit sectors, BDPs should be developed. An effective disaster management system requires collecting baseline data that are comprehensive, accurate, timely, and accessible.* Without these data points, real-time situational awareness is impossible. To build baseline data points, you should look across municipal departments, nongovernmental organizations (NGOs), and corporate entities.

In this chapter, we describe how to implement the intelligence cycle in emergency management, discuss situational awareness in emergency management, and discuss domestic intelligence data points. We call these data points baseline data points (BDPs). These BDPs are community indicators and provide insight into the preexisting conditions in a community. First, we understand that emergency managers are not at the top of an organization chart. In order to implement social intelligence, buy-in from elected leaders, city managers, and other government executives is essential. The first section in this chapter will focus on securing buy-in from management and developing an implementation strategy.

By the end of this chapter, you will know how to implement the intelligence cycle in emergency management, and understand the connection between the intelligence cycle and situational awareness in emergency management, what BDPs are, where to find them, and how to harness the information embedded in the BDPs to create a socially intelligent emergency management system for your community.

IMPLEMENTING THE INTELLIGENCE CYCLE IN EMERGENCY MANAGEMENT

In Section II, we discussed the various stages in the intelligence cycle. They are planning, collection, processing, analysis, and dissemination.

---

* Debra F. Laefer, Alison Koss, and Anu Pradhan (2006). "The Need for Baseline Data Characteristics for GIS-Based Disaster Management Systems." *J. Urban Plann. Dev., 132*(3), 115–119.

In this section we discuss how to adapt the intelligence cycle for social intelligence.

- Planning: We know now that perceptions of communities and people drive policies and programs predisaster. We also know that many of the preexisting conditions in a community are rooted in historical and economic issues. In order to develop a social intelligence cycle, it is important for emergency managers to ensure the end product is a result of an objective analysis. It is also important to recognize that much of social policy is rooted in subjective interpretations of who is deserving and undeserving. For social intelligence to work, that approach must be divorced from the analytical framework. Instead, this subjective thinking must be taken into account as social intelligence is gathered, analyzed, and disseminated. In sum, social intelligence must create a framework to objectively analyze issues related to social vulnerability, but must also factor in how subjectivity bleeds into the social construction of issues. The planning phase is considered the beginning and end of the cycle. As such, the historical frames, subjective understandings of vulnerability, socially constructed views, and other structural factors must be continually studied both prior to the collection of social intelligence and again once the intelligence has been compiled.
- Collection: The collection of social intelligence will rely on qualitative information, quantitative data, and other information that provides emergency managers with enough information and context to begin understanding the preexisting conditions in a community prior to disaster. Later in this chapter, we provide data points (both quantitative and qualitative) that can provide emergency managers with a baseline of how to begin collecting social intelligence and where to look for information.
- Processing: In the national security apparatus, processing intelligence involves decryption, language translations, and data reductions. This raw information is then fed into databases to be used for analysis. To implement social intelligence, emergency managers should begin to convene regular conversations with their EOC partners to access real-time information to better inform emergency management systems. In addition, breaking down information-sharing silos is critical to making a social intelligence system work. The system is only as good as the inputs. An

emergency management system is only as good as the information used to build the system.

- Analysis: In the national security apparatus, the raw data are assembled, analyzed, and then converted into actionable intelligence for use by policy and decision makers. The finished intelligence reports provide assessments of the information and its implications to the requesting authority. The same process is true for a social intelligence system. Once implemented, social intelligence can help predict how a disaster event can impact various systems and communities.
- Dissemination: The final step is the distribution of actionable social intelligence to emergency managers and policy makers. The actionable intelligence delivered is used in support of policy and decision making. Often, dissemination of actionable intelligence leads to further requirements by the consumer and, as a result, the continuation of the intelligence cycle.

The social intelligence cycle is ongoing, real-time, and never ends. If implemented, social intelligence can provide real-time situational awareness of what is happening in a community prior to disaster. The next section will discuss the need for real-time situational awareness in emergency management.

REAL-TIME SITUATIONAL AWARENESS— SOCIAL INTELLIGENCE

What is situational awareness? Intuitively, it is how you answer the following questions:

What is happening?
Why is it happening?
What will happen next?
What can I do about it?[*]

Answering these questions prior to disaster is crucial in ensuring emergency management systems are effective during and after disaster. More importantly, understanding the answers to these questions in real

---

[*] Reichenbach, Scott, Situational Awareness: Key to Emergency Response, *Fire Engineering* 162(3): 137–140, 2009. Academic Search Complete, EBSCOhost (accessed May 18, 2014).

time is most important. The ability of emergency management to take action is based on the quality of information it has access to. Unfortunately, many emergency management plans and systems build very static plans—there are annual updates, and much of the planning process only encompasses accounts for life safety operations. The shortcoming of this approach is that disasters impact communities very differently. The pre-existing conditions drive how communities are impacted. So creating a very flat emergency management system ignores how communities ebb and flow predisaster, and how those ebbs and flows manifest themselves during disaster and post-disaster.

Implementing social intelligence will help emergency managers understand their communities in real time. Strategies to ensure real-time situational awareness include

1. Developing and maintaining relationships with government stakeholders (internal and external), the private sector, and the nonprofit sector.
2. Creating a framework to develop an acute understanding of historical context related to prior disaster events.
3. Creating a framework to divorce the subjectivity associated with social vulnerability and view social vulnerability issues just for what they are. Simply put, creating an environment to provide a cold analysis of the situation.
4. Ensuring there is continued assessment of the elements of the intelligence cycle. That is, understanding that data collected today may not be relevant in the future. Emergency managers should develop a system to assess data points and the source points.

Maintaining real-time situational awareness requires emergency management systems to maintain a very objective perspective on social conditions in a community. This might sound like we are advocating that you ignore the conventional attitudes related to social vulnerability. We are. Our point is to create a picture of what is really happening in a community instead of what people believe is happening based on their subjective frames on public policy.

CONNECTING THE DOTS

The previous section discussed the importance of real-time situational awareness rooted in the social intelligence framework. Now, the question

132

is what to do with the information. Before we provide the information architecture for social intelligence (the information emergency managers should collect), we would like to discuss the importance of developing a strategy to make connections with the information.

> It was inevitable that as the months passed after Sept. 11, reports, memos and speculations would be found that, in retrospect, would seem to have provided early warning—if only someone had connected the dots. While some pre-9/11 items of intelligence today seem like red flags, pulling together incomplete or ambiguous fragments of information into a credible and compelling analysis is more difficult than the Monday-morning quarterbacks would have you think. Especially doing so convincingly enough to prompt high-level, high-risk decisions.*
>
> A key problem prior to Sept. 11 was structural. Since 1986, representatives of a number of national security organizations and the FBI have worked together daily in the CIA's Counterterrorism Center, where information from abroad is shared, integrated, analyzed and acted upon. Before Sept. 11, there was no comparable formal organization for working-level contact among the domestic agencies of government—or between them and the national security agencies. While there appear to have been a few dots to connect, there was no effective mechanism for those connecting lines to cross domestic and national security boundaries.†

We cited this section from Secretary Gates's interview with *Time* magazine to illustrate the importance of building relationships across departments and agencies, creating a system to share information, and then using these relationships and systems to make sense of what the information presents. This is no small task. But in order for social intelligence to work, connecting the dots on issues are the emerge predisaster can help emergency managers better focus resources during and after disaster.

To connect the dots, emergency managers should do the following prior to implementing social intelligence:

1. Ensure there is buy-in from leadership. This means ensuring there is buy-in from elected leaders and department leaders.
2. Create working groups to explain the rationale and philosophy behind social intelligence.

---

* Gates, Robert M., A Former CIA Chief on "Connecting the Dots," *Time* 159(21): 30, 2002. Academic Search Complete, EBSCOhost (accessed May 18, 2014).
† Gates, Robert M., A Former CIA Chief on "Connecting the Dots," *Time* 159(21): 30, 2002. Academic Search Complete, EBSCOhost (accessed May 18, 2014).

3. Develop relationships across working groups to lift the veil on organizational silos.
4. Develop communications and systems to share information across an organization.
5. Use social intelligence BDPs to develop easy-to-update dashboards. Develop key dashboard indicators based on quantitative BDPs. Examples include current stock of available affordable housing, number of units in demand, number of food pantries, etc.
6. Finally, understand there is always a trail of clues after the fact. Use the Monday morning quarterback example to backtrack after a disaster response and recovery. Use lessons learned and historical information to add value to the social intelligence system.

Here are some real examples of connecting the dots:

1. Many commercial fishermen were recovering from Hurricanes Katrina and Rita when *Deepwater H*orizon spilled millions of barrels of oil into the Gulf. Many commercial fishermen had recently purchased fishing vessels from recovery funds post Katrina and Rita. *Deepwater Horizon* created an impossible situation for commercial fisherman who had to make payments on new vessels.
   • Connecting the dots: While BP Amoco eventually set up a $20 billion recovery facility, the immediate impacts felt by fishermen could not be mitigated without a more real-time response. The commercial fishing industry was also supported by fish processing plants, sports fishing, and local tourism and hospitality. The interconnectedness of the local economy via the Gulf created a cascade of impacts across the Gulf Coast.
2. During the Hurricane Andrew recovery efforts, many households did not qualify for emergency housing. Why? The Federal Emergency Management Agency (FEMA) policies created a preference for nuclear families.
   • Connecting the dots: Emergency management and FEMA could have been more flexible with the interpretation of the policy. Allowing all families to access emergency housing could have prevented issues from arising elsewhere in the long-term recovery operation.
3. During the Chicago heat wave, a majority of the deaths occurred in African American neighborhoods.
   • Connecting the dots: Because of widespread power outages, many residents were without air conditioning. This forced

people outside. In turn, people turned on fire hydrants. The number of open hydrants caused a loss in water pressure in people's homes. Emergency generators and cooling centers could have prevented the cascade of impacts in African American communities.

The situations discussed above are basic examples of connecting the dots. With a robust social intelligence system, emergency managers and policy makers can drill down on preexisting conditions to better understand social isolation, concentrated poverty, underground economies, etc. Just as important, emergency managers can begin to understand the inherent strengths built in to communities—strong social networks, parish systems, underground economies, common-law trades, etc. It is important to understand that even vulnerable communities have strengths. Harnessing those strengths requires emergency managers to collect social intelligence and then connect the dots.

DEVELOPING BDPS—MANAGEMENT PERSPECTIVES

In order for social intelligence (SI) to work in your community, it is critical to develop a strategy for implementation. Social intelligence is not a one-size-fits-all strategy. Each community will use elements of SI as needed because every community is different.

A SI implementation strategy should include the following:

- Ensure there is buy-in from elected officials and administrators.
- Conduct an historical analysis of major events—the impacts to infrastructure and social systems.
- Examine from census and other information the social strata and demographics in a given community you represent.

Buy-In from Elected Officials and Administrators

For SI to be implemented in a community, buy-in from elected leaders and administrators is paramount. Like any initiative, management must drive its implementation and continued participation. Given the construct of local government hierarchy, we recognize that emergency managers will not drive the implementation. Emergency managers can help foster the conversation needed to implement SI.

We also recognize what the pushback from emergency managers will be

- Implementing SI is time- and resource-intensive.
- Most emergency management offices are small and understaffed and cannot handle the additional responsibilities of SI.

The pushback is exactly why high-level buy-in is important. All levels of government continue to spend money on disaster response and recovery without understanding how to best target funds and resources. SI implementation will help change that.

Next comes the issue of how much information you should collect and how much is enough. Obviously, the more information one has, the more options there are. Volumes of information can also turn into noise and cloud decision making.

How much information do you collect? How deep down do you drill? The answer is really up to the policy makers and emergency managers, as well as how much information is easily or readily available. Getting the buy-in from leadership should involve a series of discussions—they should understand that most issues that come up post-disaster have nothing to do with the disaster. Leaders should understand that by mapping out the connections between government, corporate entities, nonprofit services, and social systems, they can begin to connect the dots. By connecting the dots, leaders can then mitigate problems before they begin. You learned about this way of thinking in Section II of this book. Once you have buy-in, you are ready to develop a social intelligence implementation strategy (SIIS).

Convene Regular EOC Conversations

In order to develop a SIIS, emergency managers should convene elected leaders, city/town managers, and department heads in the same way they would during an Emergency Operations Center (EOC) activation. A series of discussions should be held on issues such as revenues, service structure, and community development. Emergency managers should also reach out to NGOs and major employers. A good place to start is to call all EOC partners. These are the organizations that emergency managers normally rely on during and after a disaster event. These organizations can provide emergency managers with information regarding social service provisions and employment data/trends. Based on these discussions, emergency managers should be able to develop a baseline level of jurisdictional competence. Government services, social service provision, businesses, and people are all interconnected. This means a change to social service provision may impact tax revenue collection. A change to government service may require more social service provision. And a change to the economy will have a ripple effect on everything mentioned above. These connections must be understood prior to disaster because they will become major issues after disaster. Convening regular EOC conversations (and outside mandated emergency exercises) not only shows the interdependencies in communities, but also helps break down silos. SI depends on accessing information across a variety of entities in a community.

UNDERSTAND THAT INFORMATION IS CURRENCY FOR EMERGENCY MANAGEMENT

Most organizations use information as currency. This is certainly true in the public sector. Interdepartmental cooperation is often stymied because of a competition for scarce financial resources. For emergency management, ensuring departments are talking, sharing, and working collaboratively prior to disaster will ensure emergency response and recovery efforts will flow smoothly. Laying the foundation for increased communication and information sharing, emergency managers can break down silos.

Breaking down information silos will lead to other successes for emergency managers. For example, for SI to work, socializing the philosophy of increased cooperation and collaboration among departments, policy makers, and elected officials is vital.

HISTORICAL ANALYSIS OF MAJOR EVENTS

For SI to work, emergency managers must understand how their communities ebb and flow prior to disaster. A historical analysis of major events and disaster events is necessary. For example, if the major employer in your community is a major auto parts supplier, any changes to the auto industry at a global level would impact your community. So if the major employer downsized 5% of its workforce, it would be important to understand how this downsizing would impact the tax base, tax revenue collections, and social service agencies. In addition, emergency managers should conduct an historical analysis of disasters that have impacted the jurisdiction. That analysis may exist in the form of a hazard vulnerability analysis. This document should be shared with all EOC partners. An all-hazards thinking is paramount, but EOC partners should also be aware of disaster trends and other major events impacting their jurisdiction.

Again, you may wonder why emergency managers should be concerned with events not considered to be disasters. Let's take the example of the auto parts plant. If the plant lays off 5% of its workforce, there will be an immediate impact to sales taxes in town. If the layoffs are long-term or permanent, it is likely property tax collection rates will decrease. Foreclosures will increase. Capital improvement projects will be impacted. Public safety operations will be impacted. Now add a disaster to the mix. All those preexisting conditions will have to be considered in addition to the impacts from the disaster. Knowing the history of the jurisdiction prior to disaster can help predict what may happen in the future. It is also a useful exercise to understand the community's dependencies across the public, nonprofit, and private sectors.

If a historical analysis or vulnerability analysis does not exist, it is important for emergency managers to develop such a document. By studying various disasters (as shown in Chapter 2), we found that there are lessons learned and trends that emerge. We studied various disasters across the country over the course of a century. We found that affordable housing, evacuation, poverty, single-sector communities, and institutionalized racism are just some of the issues that emergency managers contend with. More importantly, these impact emergency management efforts after every disaster. Why? It goes back to preexisting conditions in a community.

To ensure the analysis is complete, emergency managers should request information on major events impacting their community from the following sources:

- Internal emergency management information
- Departments within government
- Units of government—county, state, federal agencies
- Local Red Cross
- State emergency management agency
- FEMA
- Local and regional NGOs
- Chambers of commerce
- Major businesses/employers identified in municipalities' certified annual financial report (CAFR)
- Local historical societies

- Local religious institutions—to understand generational impacts of disasters
- Newspapers and other media sources
- Local post offices/postmasters

Obtaining this information will also help emergency management officials build key relationships. More importantly, the entities and people listed above should talk and engage one another. They are key partners in major events and disasters. There is no reason they should only engage one another during a crisis.

Finally, once you have built relationships with the various EOC partners and other organizations to help create a historical analysis of major events impacting your jurisdiction, ask more questions. Local churches, NGOs, and other agencies will have a wealth of information related to disasters. The local Red Cross will have data on individuals and families requiring shelters and housing assistance after previous disaster events. Use this information to learn about how populations have been impacted by different types of disasters.

Leveraging the strategies discussed in this section will help emergency managers develop the baseline data points required to increase jurisdictional competence. By developing mechanisms to collect and analyze BDPs, emergency managers can start utilizing social intelligence to increase jurisdictional competence. In the next section, we have outlined the various BDPs that can be used to analyze social intelligence. These BDPs detail a potential conceptual framework for developing a SIIS. After SIIS implementation, a socially intelligent architecture will emerge (BDPs).

BASELINE DATA POINTS

In this section, we lay out basic BDPs that are applicable to any community or jurisdiction. These baseline BDPs can serve as the foundation for developing a social intelligence infrastructure in your community.

The baseline BDPs are broken out by municipal partners, nongovernmental entities, and the private sector. Next to each BDP is a listing of quantitative and qualitative data points an emergency manager will need to access to implement SI and ensure preexisting conditions are understood prior to disaster.

Municipal

- **Police:** Crime statistics, recidivism rates, police staffing strength, new hires, and attrition rate.
- **Fire:** Types of apparatus, capacity, daytime/nighttime population variations, locations of hydrants.
- **Public works:** Infrastructure by function, construction updates, aboveground and underground infrastructure, age of infrastructure, green infrastructure.
- **Administration:** Tax collection rate, general fund surplus, long-term debt obligations, property value trends, demographics, renters/owners, tax base breakout (corporate/retail).
- **Building:** New developments (type), existing residential development ratios (single family vs. multifamily), commercial real estate buildings and ratios (low and high density), code trends.
- **Finance:** Fund balance, budget trends, tax collection by type and dependency, pension fund status, other post-employment benefit liabilities, bond ratings, ability to borrow, capital improvement plan and funding.
- **Education:** Number of students, number of schools (primary, secondary, and high schools), number of buildings, dropout rates.
- **Hospitals:** Uninsured cases, public health data, emerging trends.

Nongovernmental Entities

- **Immigrant services:** Estimate of undocumented population, documented immigrant populations, support needs.
- **Homeless shelters/public housing/supportive housing:** Bed counts.
- **Senior residences/assisted living:** Number of seniors in a community, estimate of how many are living on fixed incomes, number of senior residential facilities.
- **Poverty reduction agencies:** Poverty trends, percent of population living at poverty line, percent of population living at 100% below and 200% below poverty line, Temporary Assistance for Needy Families (TANF) recipients.
- **Red Cross:** Available beds and emergency supplies.
- **Universities:** Number of students, residential population, commuter population, dining facility capacity, dormitory capacity, weekly update number of students living on campus, number

of staff and faculty, number of hourly employees, number of research animals, lab functions and materials.
- **Hospitals:** Beds, trauma beds, surgical capacity, primary care capacity, surge capacity, isolation capacity.

Corporate

For this set of BDPs, it is important to access the community's certified annual financial report and work with the municipal finance department and local business organizations and chambers of commerce. The purpose of this category of BDPs is to understand how major employers and the local tax base influence how local government funds services and programs.

- Revenues
- Philanthropic capacity
- Hourly employees
- Salaried employees
- Number above minimum wage
- Supply chain dependence

DRILLING DOWN ON BASELINE DATA POINTS

Baseline data points = Social intelligence = Jurisdictional competence

The previous section broke out the three categories of BDPs. We also provided some basic qualitative and quantitative measurements for each BDP.

In Section II of the book, we learned about situational awareness and emergency management. This section will provide emergency managers, policy makers, and elected officials with a comprehensive understanding of how BDPs can be woven together to build on existing jurisdictional competence. The examples are not comprehensive. They are supposed to provide you with a conceptual framework for developing BDPs specific to your community. The following sections will provide you with a description of why the BDP is necessary and then provide you with the strategic rationale for building the BDP specific to your community. Then we will provide some examples of how to leverage the BDPs with cross-stream, upstream, and downstream partners.

Municipal BDPs

Police: The police BDPs are useful in understanding how police protection impacts a given jurisdiction. Crime rates, recidivism rates, and police coverage directly impact the well-being of a community. Issues related to public safety can have a dramatic impact on property values, population growth/decline, and in the classroom.

Preparedness: When a jurisdiction is developing emergency plans, police BDPs play a significant role. Police coverage by jurisdiction, community, and beat can have major impacts on how a community prepares for disaster. Understanding how police manpower is deployed can give emergency managers a picture of what is happening across their community. A jurisdiction collects revenues from multiple sources, and police protection is often a major expenditure. Revenue ebbs and flows, revenue projections, and other economic trends impact how a jurisdiction polices itself. The police BDP should be factored with other BDPs to identify areas of opportunity or concern. This information then should be factored into an emergency plan.

Mitigation: We often hear that an ounce of prevention is worth a pound of cure. When we think of policing, we think of arrests and incarceration. Crime prevention is funded, discussed, and studied, and yet we still don't have many answers. Crime is not just caused by a criminal action—it is often the cumulative impact of structural factors within a jurisdiction. High rates of poverty, low-quality schools, and safe passage are just a few factors that contribute to high crime rates. High crime rates are not just an urban issue. In rural America, we are seeing methamphetamine and prescription drug epidemics. Drug cartels are now targeting higher-income neighborhoods as distribution centers for narcotics. Mitigating crime involves attacking the structural factors that incubate crime. To attack those factors, we must collectively, as a society and as communities, address poverty, education, and affordable housing.

For example, in 2005, it cost New York City $68,000 per year to house an inmate on Rikers Island, while residential drug treatment costs $17,000 annually.* Our nation's jails are filled

---

* *Inside Rikers: Stories From the World's Largest Penal Colony.* Reviewed by Mark Radosta, L.I.C.S.W.

with repeat drug offenders, and incarceration seems to be an ineffective deterrent. Cost savings could be generated by mitigating incarcerations via community-based drug treatment in high-crime/high-drug-arrest areas. The cost savings could then be diverted to schools, afterschool programs, safe passage, and poverty reduction. For emergency management this information is crucial; it can help policy makers in the field understand what contributes to preexisting vulnerabilities. These same conditions will impact a response and recovery. Working with SI partners to understand and potentially mitigate issues prior to disaster can help communities better allocate resources prior to disaster.

Response: Police forces make up the first responder response during disasters. Arming police officers with more information increases situational awareness. Understanding the composition of police departments can help emergency managers allocate personnel. Police are integral components of evacuations. Providing police officers with other BDPs can help them with overall situational awareness. In New Orleans, nearly 100,000 residents failed to evacuate due to a lack of personal resources (car, cash, and credit). The police department knew about concentrated pockets of poverty, but did they really understand how the structural makeup of concentrated poverty would impact evacuation? What if they had had a more complete understanding of poverty and BDPs from the social service sector? An understanding of BDPs related to poverty reduction, affordable housing, and poverty rates could have painted a clearer picture for police. See the sample BDPs below:

1. There are shortages of affordable housing everywhere.
2. Poverty reduction/public aid programs helped people survive, but did not help them get ahead.
3. Public transportation was nonexistent; car ownership and credit access were limited.

This information could have aided in the design of rescue efforts. Increased situational awareness during the response phase improves recovery efforts.

Recovery: Recovery is often the most complicated and misunderstood phase of the emergency management cycle. For police departments, recovery efforts include securing sensitive areas, neighborhoods, and protecting people from various hazards. Police officers often are asked about social service referrals, resource availabilities, and interpretation services. Police officers are often both social workers and safety personnel during the recovery phase. Arming them with more information can help build confidence in a disaster-stricken community while bolstering community relations efforts. Police officers also know their community very well, and people will often turn to them for information. By leveraging other BDPs, police officers can be better informed of available services and funding.

Fire: Like the police department, fire departments are key first responders. Firefighters play an important role across all phases in emergency management. The fire BDP aims to understand the types of apparatus, capacity, daytime/nighttime population variations, and locations of hydrants. These baseline data points can provide emergency managers with an understanding of fire assets.

Preparedness: Fire departments spend almost all of their time preparing for disaster. They respond to many calls, but most of their time is downtime. So they are constantly maintaining their equipment and surveying the landscape for emerging trends. Jurisdictions are in constant motion—people move in and out, and revenues increase and decrease. These shifts have impacts on capital purchases of the fire department. When population density increases, fire apparatus upgrades may be necessary. If there are vast differences in population from daytime to nighttime, service structures will change. These changes impact how a jurisdiction prepares for disaster.

Mitigation: Fire departments work closely with building departments to make recommendations to municipal code. They share information on emerging trends. This information can be beneficial to other municipal partners, including those making policies related to zoning and building. This information should be shared via the fire BDPs.

Response: Firefighters are heavily involved during disaster response. Their expertise and assets are utilized to rescue people, put out fires, conduct fire safety equipment tests and drills, and inspect the structural integrity of buildings. The information learned during response can aid other municipal departments.

Recovery: During recovery efforts, firefighters can help inform city managers and building departments of lessons learned regarding structures. This information can help inform reconstruction efforts.

Public works: The public works BDP is an important data point. Understanding the connections between public infrastructure and investment against social vulnerability can help emergency managers better understand the nexus between community building and community investment. Public infrastructure, such as roads, sidewalks, and water mains, directly impacts municipal revenues. The backbone of a social system (neighborhood) is infrastructure. Access to public transit increases access to jobs. Properly maintained roads increase home values. This BDP can help emergency managers and policy makers make sound decisions about infrastructure investment and identify emerging trends.

Preparedness: The location, physical status, and dependencies of public works infrastructure help inform emergency managers about risks and hazards. The physical location of fire hydrants, water mains, and gas lines can help create greater situational awareness. Maintenance records and capital improvement schedules can inform first responders about the integrity of the infrastructure. City managers and building departments can make better zoning and development decisions if they know the age and integrity of public works infrastructure relative to potential demand on infrastructure.

Response: The location, physical status, and dependencies of public works infrastructure will increase first responder situational awareness. Knowledge of a gas line and its current status can help inform fire and police personnel on evacuations. Knowledge of road conditions can help inform traffic aides on reroutes/alternatives during emergencies.

Recovery: The public works BDP is critical during recovery efforts. Infrastructure is interconnected, and it is only as strong as its weakest node. During recovery, greater attention must be paid to strengthening or rebuilding weak/damaged sections of infrastructure. Recovery efforts are resource-intensive, and strengthening weak areas of infrastructure can prevent further damage. The public works BDP can also help other municipal departments prioritize recovery efforts.

Administration: Central administration (highest elected official/ city manager) is charged with making policy and service-related decisions. The administration function BDP can help inform other municipal departments of macro-level community trends. Corporate revenues, sales tax revenues, and property tax collections impact every municipal department. The administration BDP can help inform all downstream partners of emerging macro-level trends.

Preparedness: General preparedness levels will be determined by the level of concern and available dollars at the administrative level. The city manager/highest elected official has the responsibility to present a balanced budget to the community's legislative body. Priorities are developed from input from elected officials as well as the directors of various departments. By creating mechanisms through which information can flow upstream and downstream via social intelligence BDPs, overall preparedness efforts will increase and budgets will reflect social intelligence.

Mitigation: Imagine if there were a way to understand the connections between infrastructure, unemployment, and public transit. Those connections exist, and we can make them by leveraging BDPs. Administration officials can leverage data points and connect the dots. By addressing interconnected problems, you can start a domino effect. Just as disasters and economic downturns can send shock waves throughout a community, BDPs can do the same with positive results. You can mitigate preexisting vulnerabilities so that disasters cost less.

Response: BDPs can aid during response efforts. Central administration is often the source for all major policy and service decisions. The more quality information a decision maker has, the better the decision. Administration officials can also reach out to corporate and nonprofit partners to aid in response efforts. BDPs for corporate and nonprofit entities can be leveraged to create staging points, supply chain centers, and shelters. That information can then be disseminated in the EOC. Some emergency managers believe these conversations should only occur during a disaster operation. We believe that in order to better target resources during and after disaster, government and emergency management must ask the right questions prior to disaster.

Recovery: Revenue streams and demands for service are critical during recovery. Centrally, decision makers (city managers, department heads) and elected officials should understand the short- and long-term impacts of a disaster event on the community's fiscal health. To garner a robust understanding, BDPs should be in place that track a variety of data across government, private, and nonprofit sectors. For example, production capacities in the private sector may change and social service providers in the nonprofit sector may have to do more with less. Understanding these changes will lead to more informed recovery measures.

Building: The building department oversees development, zoning, and some facets of economic development. The building BDPs can help administration officials develop more holistic development plans. By leveraging the BDPs of building and public works, decision makers can understand the impacts of development against current infrastructure and future requirements.

Preparedness: The building BDPs can help inform emergency planners of potential structural vulnerabilities, buildings that are not American Disabilities Act (ADA) accessible, etc. The building and zoning BDPs can help inform zoning and building policy to reduce flooding, require stricter building standards in the event of fire/earthquakes, and disincentive new construction in areas prone to disaster.

Mitigation: For mitigation efforts, building department BDPs can help inform development standards, and green infrastructure recommendations, and help reinforce national best practices.

Response: The building department will possess architectural drawings, geographic information system (GIS) renderings of various neighborhoods, and be able to work with public works to identify critical infrastructure. During a response, the building BDP can play a role in evacuations, sheltering, and recovery.

Recovery: There are many lessons learned during each phase of disaster. The building BDPs will be able to inform recovery personnel with best practices, and emerging trends, and provide historical data on zoning, permitting, and development.

Finance: Financial resources are paramount during and after disaster. A good administrator understands that the fiscal health of an organization will determine how it responds and reacts to disaster. Just as social systems are impacted by preexisting vulnerability, the financial health determines organizational resilience. You should understand how revenue streams impact the bottom line of municipal departments. You should always be aware of how your response and recovery partners are faring. You can then use that information to build more robust plans.

Preparedness: A local government must produce an annual balanced budget. After the fiscal year has ended, the budget and financial statements are audited by external auditors. They produce a document called the certified annual financial report (CAFR). Both the annual budget and CAFR contain a wealth of data that should be plugged into the financial BDP. Tax revenue collection, collection percentages, interfund transfers, fund balances, outstanding debt, and personnel costs are all important pieces of information. If you look at multiple years of budgets and CAFRs, you may spot trends. Then, if you examine real-time data like monthly receipts, fees, and fines, and then study employment, crime, recidivism, and property sales figures, you will be able to spot trends. When historical data and real-time data are combined, you are socially intelligent regarding financial health. This will help you maintain your emergency plans in real time.

Mitigation: You can leverage the finance BDP to help mitigate the circumstances that stress municipal departments. For example, if there is an uptick in crime or drug activity, the finance and police BDPs can be leveraged to project impacts of additional police hires, drug programs, and equipment purchases. BDPs from the corporate and nonprofit sectors can be isolated and then entered into algorithms to determine resource allocation effectiveness.

Response: Municipalities are struggling to fund services because of declining revenues and unfunded pension obligations. Response operations will be based on fiscal health. To ensure efforts are adequate, finding efficiencies is important. By connecting the dots (BDPs) across the entire landscape, finance directors and administrators will be able to allocate resources more effectively.

Recovery: Recovery efforts may be partially funded by state and federal dollars. The proper allocation of those dollars can only be achieved if there is an intelligent understanding of the community. Jurisdictional competence is necessary when dollars flow in—the dollars must fund recovery efforts holistically.

Education: Education data may not seem relevant for emergency planning but they are. Education data can inform administrators and emergency managers of jurisdictional capital. School performance says so much about a community or neighborhood. It can shed light on property values, infrastructure, access to banks, and other crucial data. Education systems are often the bedrock of a community. The data can inform administrators and emergency managers of the future. Schools are also an indication of government investment. In cities across America, school districts are closing schools in favor of charter schools or to align schools with changes in population. While some of these actions are justified, in most cases, especially in very poor communities, public schools are the last vestiges of government investment. Closing a school can lead to the perception that the local government has given up on a community. School closings send messages to companies and people seeking to make investments in a community.

Preparedness: School data like number of students, number of schools (primary, secondary, and high schools), and number of buildings can help you build emergency plans.

Mitigation: One of the data points not included in preparedness is dropout rates. Dropout rates and school performance can be linked to job losses, poverty rate increases, and general economic decline. An increase in dropouts can be used as a trigger—once activated, you can inform social service providers, transitional job centers, and police departments of the increase. They can then be prepared for surges in service demands or requests.

Response: Understanding the connections between number of students, number of students that require free lunches, number of schools, and number of facilities can help ensure robust emergency response operations.

Recovery: After a catastrophic disaster, complete jurisdictional competence is required. If you are sending disaster evacuees to other cities, emergency managers should provide host cities with education BDPs. They will be taking on your populations, and good information is required to help recovery and resettlement.

Hospitals: There are still local governments that provide healthcare. These community hospitals are a key partner in any emergency situation. The service they provide is directly impacted by public health issues, increases/decreases in poverty and employment, health insurance coverage, and other BDPs.

Preparedness: Hospitals play a central role in emergency management. They deal with the day-to-day emergencies and then triage medical services during and after disaster. Hospitals track gunshot wounds, public health information, and other health-related data points. This information should be aggregated and entered into a hospital BDP. This BDP can then inform municipal partners of emerging trends.

Mitigation: Hospital BDPs can inform many other BDPs across government, private, and nonprofit sectors. Information collected at hospitals can inform EOC partners of emerging public health trends linked to obesity and poverty. In Chicago, obesity and poverty are linked to a lack of primary care access and food desserts. Hospitals in Chicago see the connections between poverty, food, obesity, and major health problems. They should be in the loop as policy prescriptions are developed. Hospitals may also start to see warning signs before other EOC partners—they can then pass on data related to emerging trends and mitigate issues before they get worse.

Response: Hospital BDPs can play a significant role in any disaster response. Disasters may cause casualties, and having information regarding triage, trauma capabilities, isolation centers, and medical surge will help emergency managers plan for different types of disasters. Understanding the capacities and surge capacities of local hospitals can create opportunities to develop mutual aid agreements with other hospitals.

Recovery: After recovery efforts commence, populations may be in need of mental health services and primary care. Hospitals' data could inform policy makers and administrators of how to effectively provide medical services to affected populations. Hospitals could also track public health issues post-disaster so best practices can be gleaned from current practices.

NONGOVERNMENTAL ENTITY BDPS

Nongovernmental organizations (NGOs) have increasingly played larger roles in emergency management. Government often provides assistance in the immediate aftermath of disaster. Until there are state and federal declarations, disaster victims are on their own. In fact, even after disaster declarations, bureaucratic nightmares await disaster victims. *Emergency managers need to realize that the time period between a disaster and disaster declaration is critical.* A small flood that goes unchecked can lead to massive mold and structural damage. A family without a safety net can go from living just above the poverty line to being destitute. A senior citizen without access to medications could see medical problems unfold in a matter of days. Disaster mitigation recognizes that disasters will occur and cause damage, but strengthening infrastructure is one way to mitigate massive damage. That same logic should apply to social systems. Social policy is intertwined with politics and public perception. You have already read about the social welfare policy and disaster policy—both are mired in ideological debates about deserving and undeserving populations.

How can we help shape different perceptions? How can emergency managers help government and people understand that disasters occur when there is an impact on humans? It begins by understanding that NGOs provide service to people when government does not. It is the NGOs that fill that void because sometimes state and federal money never flows in. These organizations are the subject matter experts on the human condition. But that isn't to say that emergency managers can't also incorporate social aspects into their planning, in addition to the infrastructure. They work with every type of population, and they understand how needs and resources vary. The information that NGOs collect is valuable. This information is often overlooked and is often the primary reason emergency plans look like templates. The plans do not reflect what is going on on the ground, in the daily lives of people.

This section will provide you with insight on the types of information NGO BDPs can provide emergency managers. This information can help inform emergency plans, regional catastrophic plans, and continuity of government plans, as well as help administrators make better decisions about budgets and resource allocations.

Immigrant services: Organizations that provide services to immigrants, refugees, and people seeking asylum perform an important social service function. They are charged with helping individuals and families with a comprehensive suite of services.

151

Services include English language training, job skills training, career development, culture training, and assisting in navigating public welfare programs.

Preparedness: In recent years, emergency managers have recognized the need for culturally competent messaging. Disaster preparedness measures and evacuations are now increasingly focused on disseminating information through various languages and in a culturally competent manner. Emergency managers can use this BDP to understand population shifts and help administrators make decisions on resource programming.

Mitigation: Mitigating the impacts of disaster requires getting the message out to people. Immigrant community service providers represent significant percentages of a jurisdiction's population. These organizations can be leveraged to disseminate information. The information collected locally by these organizations can inform emergency managers and administrators of emerging trends in employment, poverty, etc.

Response: Immigrant services can be leveraged to help expedite evacuations as well as inform emergency managers of access to resources, support requirements, and language requirements.

Recovery: The immigrant services BDPs can inform administrators of the types of services required by various communities. The information can ensure that recovery processes are reflective of immigrant population needs and concerns. Immigrant service providers are in the business of providing basic needs to populations. The models for resettlement can be a template for resettling disaster victims.

Homeless shelters/public housing/supportive housing: Housing is important during every phase of emergency management. Access to affordable housing is a problem in every metropolitan city. It is also a problem in medium-sized and rural communities. Homeless shelters are a critical resource during everyday life—after disaster, they can aid in sheltering disaster victims.

Preparedness, mitigation, response, and recovery: Bed counts—they can vary daily, and understanding what a jurisdiction has available on any given day is critical information. Smaller disasters like extreme cold or heat can trigger the need for bed count information. You may recall the winter storm that hit the New York–New Jersey area immediately after Hurricane Sandy, compounding the disaster for those whose homes were otherwise destroyed or rendered unlivable. This information should be updated daily. This critical piece of intelligence can help emergency managers understand need during and after disaster. Jurisdictions can either receive or send out disaster victims. Available beds vary daily and are often very limited. Affordable housing is also very limited. This information should be available to decision makers prior to accepting or sending out disaster victims.

Poverty reduction agencies: There are many social service and advocacy shops that specialize in providing poverty reduction programs. These service providers study vulnerability, poverty, credit access, and many other issues.

Preparedness*:* While developing emergency plans, emergency managers can tap in to data related to poverty trends, percentages of populations living at various levels of poverty, and TANF recipients. These data will help inform emergency managers and administrators of any tectonic changes in the socioeconomic landscape.

Mitigation: BDP data regarding poverty can be connected with public works, crime, and building data to understand the structural causes of poverty. By mapping out the structural causes, tipping points for change may be identified. These tipping points may not be achievable in the near term, but can serve as guideposts for new policies and programs.

Response: By understanding the nature of poverty within your jurisdiction, you can prevent another Katrina-type response failure. You can advocate for evacuation resources or target disaster-related information more appropriately.

Recovery: You read about the Heartland Alliance approach post-Katrina. They were focused on rebuilding livelihood. After disaster, recovery is a process that goes beyond 72 hours. People may need help finding work, navigating public assistance programs, locating housing, finding schools for their children, and accessing mental health services. Understanding the level of need is necessary. To quantify need, emergency managers must be jurisdictionally competent.

Red Cross: The Red Cross is a valuable information resource. It already provides assistance to individuals and families post-disaster for up to 72 hours. Its resources are limited, and it cannot extend services beyond that threshold. That said, it can be leveraged to extract data on resource needs beyond 72 hours.

Preparedness: The Red Cross maintains a steady stock of beds and other supplies. This stock fluctuates daily. These data should be captured and fed into a social intelligence BDP.

Mitigation: Red Cross information and BDP data can help inform emergency managers and administrators on disaster impacts.

Response: The Red Cross is already involved in many response operations. Their BDP data should be shared in advance of disaster.

Recovery: The Red Cross provides important recovery services—their BDPs and expertise should be leveraged when possible.

Universities: Colleges and universities are mini-cities within a jurisdiction. Many maintain their own emergency management offices and police departments. The resources that universities and colleges can bring to bear during and after a disaster often go untapped. Institutions of higher education are filled with administrators and educators that are subject matter experts in a number of areas that could better inform emergency planning. As an example, many North Dakota State University students have volunteered to help sandbag or otherwise prepare for Red River flooding that has occurred in recent years.

Preparedness: Universities and colleges that have research units will need special attention when developing/updating emergency plans. Hazardous materials, national security-sensitive information, and research animals are all elements of research programs. Universities and colleges also house and feed many students. The needs of student communities must be accounted for in emergency plans. Information related to dormitories, bed counts, food services, and research is an important component of a jurisdiction's emergency plan.

Mitigation: Research and public policy innovations are products of universities and colleges. Faculty and students are continually finding ways to improve systems and public policy. Emergency managers should leverage these resources to innovate emergency management mitigation programs.

Response: Universities can aid during response efforts by providing critical data—number of students living on campus, research facility information, and information related to hazardous materials.

Recovery: Once life safety operations have concluded, universities/colleges can help emergency managers by providing beds, food service assistance, and other logistical support.

CORPORATE BDPS

Investors demand almost real-time information about the companies they invest in. This same logic must apply to how municipalities survey the jurisdictional landscape. If Company A logged $100 million in revenue in 2010, there is no guarantee that in 2011 revenues would grow. The global market may change such that revenues may shrink or grow exponentially. Administrators and emergency managers should be aware of how economic changes impact the jurisdiction's bottom line. The corporate sector is not only a revenue engine for local government but also a litmus test of the overall health of the jurisdiction. BDPs that can help inform administrators and emergency managers include revenue, revenue growth, hourly employees, salaried employees, number of employees above minimum wage, and supply chain dependence. These BDPs are critical during all phases of emergency management. Why? Well, let's suppose a disaster shuts down the major employers in your jurisdiction. Until any aid arrives, the workers from those companies will rely on their social safety net. For many hourly employees and those making hourly wages just above minimum wage, a slight disruption in wage accruals can cause a spiral into poverty. Understanding this information can help emergency managers coordinate preparedness programs and evacuations, and truly understand recovery needs.

You now have a basic understanding of how to leverage BDPs across municipal departments, nonprofits, and the corporate sector. Individually, each of these BDPs tells you little about the condition of your jurisdiction. Each BDP is like an independent gear within an engine. If you study one in isolation, you can't learn much about the entire machine. When there is a disaster, every BDP is in play. Every BDP interacts with people through a web—emergency managers should try to understand these individual BDPs and their interactions before disaster strikes.

The next section will provide you with an architecture to leverage your BDPs with existing technologies. There are a number of emergency management systems, but locally GISs are a widely used technology tool. GISs can help you map out BDPs and then help you connect the dots.

We have identified the basic BDPs to help create the framework for implementing social intelligence in your community. In Chapter 6, we outline the various information management and technology tools to leverage to provide spatial and visual understanding of how BDPs impact your community.

6

Geographic Information Systems and Modeling for Disaster

In this section we will study in depth the geographic information systems (GISs) and modeling for disasters. We will explore how GISs are used in emergency management, different software programs available, and any limitations to GISs. We will also take a look at modeling for disasters and how this helps the emergency management community plan for and respond to disasters.

A geographic information system (GIS) is "an organized collection of computer hardware and software designed to efficiently create, manipulate, analyze and display all types of geographically or spatially referenced data. A GIS allows complex spatial operations that are very difficult to do otherwise" (Pine, 1998[*]). GIS software allows users to create visual maps with the ability to display layers of information (such as roads, hospitals, geographic elements, etc.) that aid in the mitigation, planning, preparation, response, and recovery processes of emergency management. Although GIS software applications are a modern tool used widely throughout society, using GIS has been a practice since the mid-1800s. In 1854, during a citywide cholera epidemic in London, a Dr. John Snow used a map to pinpoint the source of the outbreak by analyzing data that had been collected by human sources (FEMA[†]). He was then able to trace

---

[*] Pine, John C., "Geographic Information Systems in Emergency Management." Research, Institute for Environmental Studies, Louisiana State University, Baton Rouge, LA, 1998.

[†] FEMA, IS-922 – Applications of GIS for Emergency Management. Accessed August 25, 2013. http://emilms.fema.gov/is922/GISsummary.htm.

the source of the outbreak to a well that had been contaminated. Dr. Snow took the information analysis and overlaid it on a map in order to trace the origin of the cholera outbreak. Although Dr. Snow did not have access to a computer, electronic maps, or detailed databases, he was able to utilize the information he had to find the source of the epidemic and neutralize the outbreak. Modern-day GIS applications range from creating new bus routes to modeling chemical plumes and the disastrous effects they could have on a community. GISs allow the user to

- Create or input data
- Modify data
- Store data
- Analyze data
- Output information
- Distribute data and information (FEMA, IS-922)

It is this versatility of GIS programs that has been key to private and government entities in planning for projects. GIS also supports emergency management in other functions:

- Developing and maintaining lists of GIS emergency support manpower with personnel location information, contact information, and specialized skills
- Developing lists of detailed GIS data and resource requirements to support emergency management needs
- Developing secure, redundant GIS layers of local city/county critical infrastructure data, including a DVD set of critical data with integrated data viewing and printing applications (FEMA, IS-922)

GIS COMPONENTS

Now let's take a look at the basic properties of GIS programs. GISs consist of three basic components: base maps, input data, and analysis methods and tools (FEMA, IS-922). Base maps are a combination of spatial data that have been organized into layers that can be manipulated to create the desired map. Layers can be added or removed based on what information the user wants to use. Types of base map features include aerial photography, water features, man-made boundaries, and other geographic and environmental data. It is important to note that most of this information is readily available to all jurisdictions through regional vendors.

The next component of GISs is input data. Input data are categorized into two groups: raster data and vector data.

- Raster data are made up of thousands of individual pixels merged to create a digital image (e.g., a photograph). The smaller the pixel, the higher or more detailed the resolution of the image becomes.
- Vector data include the points, lines, and polygons that represent the locations and boundaries of map features. Vector data would be used for
  - Mapping streams
  - Showing detailed twists and turns
  - Depicting roads or utilities
  - Identifying exact intersections
  - Pinpointing the exact location of an elevation measurement (FEMA)

Within the two categories, input data is classified into three types: point data, aggregate data, and line data. Point data are anything that can be clearly identified on a map. This includes various types of information, such as locations on infrastructure, vital equipment, and other assets. Aggregate data are made up of large amounts of information combined to describe a certain geographic area. Aggregate data are typically used to protect individuals' privacy during disaster activities. Line data include information such as streams, roadways, and jurisdictional boundaries.

The final component of GISs is analysis methods and tools. Once the maps and various sets of data come together, users are able to use the final product to make operational decisions based on the information in front of them. As we will explore further on in the chapter, GIS is an exceptional tool available during all phases of emergency management.

GISs IN ALL PHASES OF EMERGENCY MANAGEMENT

As a professional in the field of emergency management, it is important to utilize all available resources at your disposal in order to better prepare your jurisdiction against potentially hazardous threats. With that in mind, GIS is an invaluable resource for emergency management professionals and can provide several sets of data, such as

- Storm track and damage prediction
- Wind damage prediction
- Earthquake damage prediction

- Counties that have been declared major disasters
- Demographic information for an identified area
- Road, rail, and utility locations
- Essential facilities, shelters, and other critical locations
- Repetitive losses
- Superfund locations
- Shelter locations
- Critical facility locations

By utilizing the above-mentioned data, GISs provide an invaluable tool used to support the emergency management community during all phases of emergency management. In the following sections we will explore how GISs are used during the mitigation, planning, preparedness, response, and recovery phases of emergency management.

Mitigation

During the mitigation phase emergency managers seek to reduce, lessen, or eliminate the effects of an incident on their community. Mitigation focuses on the long-term planning needs of a community in response to threats from emergency events. GISs are able to assist by identifying both the hazards and critical infrastructure present in the community. Community planners are then able to use these data to better plan locations of residential and commercial facilities in areas that lessen the chance of large-scale disasters happening. Restrictions for building in floodplains, regular inspections of critical infrastructure, and stringent building codes are all strategies that help to mitigate the catastrophic effects of large-scale disasters.

Planning

During the planning phase, emergency management personnel are tasked with developing emergency operation plans by conducting risk analysis for their community and continually working to maintain a state of readiness to respond to disasters.

GIS can assist in the planning phase by

1. Identifying and mapping natural and technological hazards:
    - Natural hazards may include
        - Earthquake faults
        - Storm surge exposure
        - Flammable vegetation

- Areas prone to severe weather events
  - Landslides
  - Floods
- Technological hazards may include
  - Hazardous materials locations
  - Transportation corridors where hazardous materials are routinely shipped (rail, highway, etc.)
  - Nuclear power plants
  - Petroleum processing and storage facilities
2. Identifying and mapping critical values at risk:
   - Population densities
   - Critical infrastructure including government facilities, hospitals, utilities, and public assemblies
   - Natural resource concerns including scarce natural resources and plant and animal habitats
3. Identifying values at risk that reside within the impacted areas of natural and/or technological hazards. GIS is used to model potential events (plumes, explosions, floods, earthquakes, etc.) and display projected areas of extreme, moderate, and light damage that could be caused by the event. Casualties can also be projected. Priorities for mitigation and emergency contingency and response plan development are highlighted through the use of GIS.
4. Developing site-specific strategies for mitigation to reduce losses; mitigation includes activities that prevent an emergency, reduce the chance of an emergency, or reduce the damaging effects of unavoidable emergencies. Mitigation activities take place before and after emergencies. Other mitigation functions may include enforcing building and fire codes, designating specific routes for hazardous materials shipments, requiring tie-downs for mobile homes, and shipping regulations for hazardous materials. Evaluate and model alternative mitigation strategies. Determine the best strategy for protecting critical assets from catastrophic damage or loss and reduce casualties. (ESRI, 2008)

During the planning phase, it is also important to schedule trainings and mock exercises for personnel, maintain current databases, and create redundant systems for a secure network. This will ensure that when an emergency situation does arise, personnel are ready and able to respond appropriately. The planning phase is very important because it allows you to explore the vulnerabilities in your jurisdiction.

Preparedness

Once the vulnerabilities in your jurisdiction have been identified by the planning phase, it is important that you address the gaps in your response and prepare for all foreseen scenarios. GIS is able to assist in your preparation in the following ways:

- Site selection for adequate evacuation shelters with consideration of where and how extensively an emergency might occur
- Selecting and modeling evacuation routes
  - Considerations for time of day
  - Considerations for road capacity versus population, direction of travel, etc.
- Identification and mapping of key tactical and strategic facilities
  - Hospitals
  - Public safety facilities
  - Suppliers to support response (food, water, equipment, building supplies, etc.)
- Training and exercises to test preparedness
  - Identify incident locations and impacts; map incident perimeters
  - Model the incident (plumes, spread, etc.)
  - Collect damage assessment, identify casualties, and prioritize for allocation of public safety resources
  - Develop and distribute incident action plans
- Providing a key capability for the command and control information system that enables situational awareness and incident management support (ESRI, 2008*)

It is also critical during the preparation phase to make sure all the databases used during the response phase are constantly maintained to enhance situational awareness. It is imperative that when you require the necessary data, they are readily and easily accessible for your staff. It is also necessary that you conduct trainings for your personnel during nonemergency periods so that when a disaster occurs, personnel know their positions and work efficiently to mitigate the effects of the disaster. The more work you can do to prepare for a disaster, the less it will impact your community.

---

* ESRI, *Geographic Information Systems Providing the Platform for Comprehensive Emergency Management* (White Paper), Redlands, CA, October 2008, 2.

Response

During the response phase, GISs are used to provide real-time updates and planning for emergency personnel. GISs allow users to see points of interest (critical infrastructure, evacuation routes, shelter locations, etc.) to assist in operational decisions. GISs also allow for the modeling of disaster-affected areas and the ability to forecast potential threats to a community. GISs support the emergency management community in the following ways:

- Provide warnings and notifications to the public and others of pending, existing, or unfolding emergencies based on the location or areas to be impacted by the incident. Areas in harm's way can be identified on the map, and mass notification can be performed from a GIS.
- Determine appropriate shelter activations based on the incident location and optimum routing for affected populations to access appropriate shelters.
- Maintain shelter location continuity of operations: supply inventories, external power requirements, shelter population capacities, etc.
- Identify the locations and capabilities of existing and mutual aid public safety resources.
- Provide facilities for the assembly of department heads to collaborate, make decisions, and develop priorities. Provide the capability to create remote connections to the command center for officials and others who need to participate but are unable to come to the command center.
- Establish the capability to collect and share information among department heads for emergency decision making to support emergency operations and sustain government operations.
- Establish the capability to share information and status with regional, state, and federal agencies.
- Support incident management operations and personnel, provide required resources, and exchange internal and external information.
- Maintain incident status and progress; facilitate damage assessment collection and analysis.
- Assure the continuity of government operations for the jurisdiction considering the impacts of the emergency.
- Prepare maps, briefs, and status reports for the executive leadership (elected officials) of the jurisdiction. (ESRI, 2008)

The use of GISs allows decision makers the ability to manipulate data in a way that takes much of the guesswork out of critical decisions.

Recovery

Even before the response phase has finished, a community must begin the recovery phase and start to rebuild from the aftermath of the disaster. It is the goal of the community to return to normalcy as soon as possible and return to predisaster function. GIS is a helpful tool during the recovery phase because it can be used to support recovery operations.

GIS is integral for recovery by providing a central information repository for assessment of damage and losses that provides:

- Identification of damage (triage based on degree of damage or complete loss). GIS allows inspectors to code parcels with the degree of damage in order to visualize specific problems as well as area trends. (GIS on mobile devices expedites the difficult damage assessment problem and can include photographs and damage reports linked to the specific geographic sites.)
  - Overall damage costs and priorities for reconstruction efforts based on appropriate local criteria
  - Locations of business and supplies necessary to support reconstruction
- Assessing overall critical infrastructure damage and determining short-term actions for the following:
  - First aid and health
  - Additional shelter needs
  - Optimum locations for public assistance
  - Alternate locations for government operations if government facilities are damaged
  - Alternate transportation routes for continued operations
  - Monitoring progress by specific location of reconstruction efforts for both long-term and short-term needs
- Publishing maps to share information with the public and other government organizations of progress toward recovery objectives (ESRI, 2008)

GIS is also used in support of

- Demobilization procedures, coordinating recovery, and restoration of unused resources
- Provision of required documentation for cost recovery to the federal and state governments
- After action reporting and subsequent planning efforts (FEMA)

In addition, communities can use aerial photography to document property losses and record the progress of a community's rebuilding.

164

GIS MODELING FOR SOCIAL INTELLIGENCE

GIS has already been leveraged by emergency managers. One good example is its utilization in the Federal Emergency Management Agency (FEMA) HAZUS-MH program. Available on the FEMA website (www.fema.gov), maps generated by the HAZUS-MH program can be downloaded by users. Examples include a GIS layer developed on the "Distribution of Utility Lifelines in Relation to Earthquake Hazard in Portland, Oregon." This map displays the water, sewer, oil, and gas lines and their potential for soil liquefaction post-earthquake.* This map can help an emergency manager understand how infrastructure is impacted by an earthquake. But more importantly, it can help authorities understand population centers in and around the most vulnerable areas and develop mitigation programs. How can social intelligence be used in conjunction with GIS modeling?

Social intelligence can help an emergency manager understand communities in real time. Communities change in ebbs and flows. So let's go back to the HAZUS-MH map of Portland. While it is important to understand how the underground infrastructure can be impacted post-earthquake, aboveground situations are just as important. We should know population densities, but also understand that density does not mean population homogeneity. From earlier chapters, you know that a wildfire will impact populations on Malibu differently than populations in Watts. So while the infrastructure underground may be homogenous at first glance, engineers understand that the infrastructure (in this case a pipe) is only as strong as its weakest section. So a weaker section of the pipe makes the whole pipeline weak. This logic applies to population centers. Dense populations that are socially vulnerable and located near weak sections of infrastructure are more vulnerable than other populations. Understanding this relationship helps emergency management become more intelligent. Similarly, if infrastructure is weakening and public works engineers are noticing an increasing trend of service disruptions or repair calls, that might be indicative of an emerging trend. That emerging trend may metastasize into a greater problem. If left unchecked, property values can start to sink and the neighborhood may change. Conversely, if poverty rates increase in a neighborhood due to corporate relocations of job losses, tax revenue collection will decrease. This decrease will make it harder for municipalities to allocate capital improvement dollars in that area, which would weaken infrastructure resiliency. The fiscal health

---

* http://www.fema.gov/library/viewRecord.do?id=3218.

of a community, the amount of jurisdictional capital, and infrastructure strength of a particular community are interconnected. They form one long chain—so the community is only as strong as its weakest link. Understanding these linkages is pivotal, as emergency managers can aid policy makers and elected officials in assessing acceptable and unacceptable risks. So how can we understand the connections between infrastructure and population centers?

The first step is to leverage existing GIS data. Municipalities are already mapping water, sewer, gas, and other utility lines. Much of the infrastructure locating has already been completed. But here are some key inputs that should be developed:

• Infrastructure and capital improvement schedule by location
• Age of current infrastructure
• Maintenance records
• Recent problems

By inputting real-time data points listed here, emergency managers can identify emerging trends (Figure 6.1). These trends can help infrastructure domain controllers work with emergency managers, city administration, finance, and elected officials to assess risk and decide whether the emerging trend presents actionable intelligence.

Then we must begin to input jurisdictional capital on a separate layer. The information used should include baseline data points (BDPs) developed for municipal, nongovernmental, and corporate entities. Emergency managers can then begin to understand how poverty rates and crime statistics impact tax revenue. That tax revenue can dramatically impact capital improvement plans for infrastructure. As that infrastructure deteriorates, property values plummet. A spike in poverty can spur a domino effect. So if a disaster impacts that neighborhood, the resources needed to respond and recover would differ from neighboring communities.

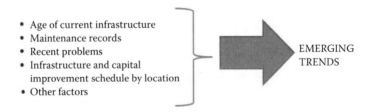

Figure 6.1   Inputting figures to determine emerging trends.

Connecting these dots with policy papers is difficult. But viewing the interconnectedness of vulnerability on a GIS map can help emergency managers look at vulnerability from a variety of perspectives. Different perspectives can help emergency managers and policy makers make intelligent decisions.

SCENARIO-BASED PLANNING

GISs cannot accurately predict in precise terms the impacts of disaster. Instead, GISs will arm emergency managers with enough information to identify the intersections across municipal departments, nongovernmental organizations (NGOs), and corporate partners. The military utilizes scenario-based planning when thinking about strategy, operations, and tactical movements. It utilizes scenario-based planning because it helps decision makers to

- Identify the complete range of plausible features
- Understand the long-term implications of trends and events
- Reveal and challenge unspoken assumptions
- Develop policies to prevent and manage adverse developments and exploit future opportunities (Weitz, 2001)[*]

These decision markers enable military commanders to fully understand the short-term and long-term implications of action and inaction. This gaming is not precise, and the variables that plug in to the various outcomes are fluid. However unstable the results, scenario-based planning ensures that the military is looking at situations from multiple angles.

> China represents the most prominent Asian military that could threaten those goals during the next 20 years. A Chinese attack on Taiwan or its use of force in the South China Sea, even if unsuccessful would disrupt East Asian commerce, heighten regional tensions, and encourage arms racing. Chinese aggression in the Pacific could compromise US alliance ties if Washington's reaction leads East Asian countries to fear either abandonment or entrapment in a conflict with China. (Weitz, 2001)

The China example from the Weitz piece is emblematic of why viewing a problem or situation from multiple angles is necessary. There are no easy answers in international relations, as actions and intent are always in question. Similarly, emergency management must also look at issues from

---

[*] All the bullet points are sourced verbatim from the Weitz article.

multiple angles. While some emergency management professionals and agencies developed scenario-based planning tools, most are not looking at issues that arise post-disaster from multiple angles.

Many municipal water departments often use water forecasting to determine water supply outputs. They base their forecasts on economic development, population growth, and pricing policy impacts (Wen, Lin, Chen, and Ming, 2007). By predicting water demand across a spectrum of variables, water department planners can identify emerging trends and anticipate necessary capital improvements and adjust future pricing schedules. For these policy makers, being blindsided with a sudden change in demand can put an entire water system at risk. But we should reiterate that this process is not always accurate. In our minds, all the scenarios can play out incorrectly—what is most important is that these planners have socialized scenario-based planning into their everyday practice.

IDENTIFYING EMERGING TRENDS

Municipalities can change in just a few short months. Our previous example of Wilmington, Ohio, comes to mind. Wilmington was a bustling middle-class community in rural Ohio. DHL was the major employer in town and a significant source of the tax revenue base. Employees purchased homes and paid property taxes. They spent their money in town and paid sales taxes, and the stores paid property and sales tax to Wilmington. Corporate taxes from DHL funneled into Wilmington's coffers. As a result, properties maintained value and the municipality was able to extend robust city services to its residents. That all changed when the economy went south in 2008. DHL began reducing its workforce, and not too long after, the town was reeling. Why?

We can begin by looking at international markets. High labor costs forced DHL to lay off staff and off-short its plant. Nationally, the economy plunged after the real estate bubble popped and resulted in the seizure of credit markets. With credit markets frozen, liquidity dried up. Banks stopped lending to companies to make good on short-term obligations. DHL likely made a strategic decision to relocate its plant by attempting to understand the confluence of international and national trends. This resulted in a local crisis. As the Wilmington DHL workforce shrunk, residents relied on savings to pay bills. After exhausting savings, they made choices between groceries and mortgages. Mortgage defaults increased, and property tax revenue sank. As a consequence, small businesses were impacted and their sales shrank, as did the municipality's sales tax revenue. You can see where we are going. A global economic change can trickle down to your community. While the thrust of the previous argument focused on negative economic impacts, the same logic can be applied with economic upticks. A new factory, housing boom, or other positive economic situation can create an influx of revenue for a community. As communities thrive, people from surrounding areas move into the community. This may trigger an increase in hospitality, tourism, and other supporting sectors. More people equals more city services. With new revenues, communities can begin to invest in infrastructure and basic services. This creates opportunity to ensure greater equity across a community and to invest in reducing social vulnerability. During these economic booms, it is critical for emergency managers to understand how economic success can be leveraged into investment to reduce social vulnerability to disaster. By implementing social intelligence, emergency managers will be able to spot opportunities for investment, and opportunities to reduce vulnerability.

As an emergency manager, you should understand that emerging trends impact your ability to plan, prepare, and respond to disaster. Without a healthy tax base, how do you plan on responding to a tornado or flood?

Identifying emerging trends is an important skill in emergency management. GIS technology can help an emergency manager visualize these trends. By visualizing these trends, you will be able to have the right conversation about mitigation and preparation. Having the right conversation before disaster strikes will ensure you will ask the right questions after disaster. Scenario-based planning via GISs can help point you in the proper direction.

By asking the right questions, you may be able to implement strategic or warning indicators into your emergency management system. These indicators may give you insight and highlight points of divergence between various scenarios (Weitz, 2001).

> For example, if one scenario of the global oil market posits a renewed energy shortage and another does not, a decline in estimated global oil reserves or higher oil futures prices would suggest the former scenario was becoming more likely. (Weitz, 2001)

So how do you embed this type of scenario planning into your emergency management program? Your program should move beyond worst-case scenario planning. The scenario planning must incorporate all the data points we've discussed, and the emergency manager must effectively communicate potential outcomes.

SOCIALIZING SCENARIO-BASED PLANNING INTO MANAGEMENT COMMUNICATIONS

As an emergency manager, you should be able to tell a story. Policy makers and elected officials must understand the potential outcomes from your scenario. The GIS technology should be leveraged to provide an accurate representation of physical assets, jurisdictional capital, and infrastructure layered against a disaster event. The narrative is your responsibility— to ensure that the outcome of your scenario most accurately represents the disaster; you should spend time with your municipal/nonmunicipal partners. They are the subject matter experts and can provide you with scenario-based outcomes. You should then piece together the various outcomes to tell a compelling story.

> Pink and Friedman both suggested that, just as manual labor was outsourced from the US and UK to overseas locations based on a low cost of labor, so are "white collar" organizational functions, going to be. As China and India produce increasing numbers of MBAs, engineers and software development specialists, the rate at which additional financial analysis, computer programming and software engineering functions will also be outsourced to these less expensive labor pools. (Chermack and J, 2006)

This scenario is compelling—it allows the reader to understand fully the ramifications of global development. It provides a complex analysis in simple form. An emergency should be able to tell the same story. To support that story, GIS technology can be leveraged to manipulate various data points across layers. So as questions are asked of you, you can then manipulate data and active warning points.

Once you are able to leverage GIS technology with robust data, you will be able to use historical data points along with real-time data to identify opportunities for improving situations.

SPATIAL DECISION SUPPORT AND GISS

As we know, disasters are characterized as high-impact/low-frequency events. While true to some extent, this thinking has cost the U.S. taxpayer billions of dollars. Why? We often underestimate the full impact of disaster on populations. To fully understand how one disaster event can have a cascading impact on populations, there is no better place to look than in New Orleans. Almost 5 years after Hurricanes Katrina and Rita, the poorest sections of New Orleans look just as they did in the aftermath of the storms.

To continue with the recovery from Katrina, we can use GIS technology to understand:

1. Where did people move to?
2. What is the state of recovery relative to new location vs. location pre-disaster?
3. How can this dynamic information be disseminated to displaced residents? (Mills and Warren)[*]

The information in the questions is relevant not only to all levels of government but also to the affected residents. It can provide a holistic understanding of how people evacuated, where they migrated to, and what the current status of their old neighborhood is. This information is a resource that increases transparency in recovery framework, and it can spark robust decisions on community rebuilding efforts, future sustainability, and resiliency measures. But unfortunately, displaced residents often had to uncover information on their own and rely on media or rumors. The information on displaced individuals could have been aggregated into information portals—this information could have then helped displaced residents work with New Orleans officials on how to structure more robust recovery efforts. But without a mechanism to start this conversation, many displaced residents started over in other cities.

To avoid situations like the one described here, emergency management must understand that there is a wealth of information that is collected, analyzed, and disseminated prior to disaster. Disaster events also produce a dearth of information during the response and recovery phases. The key is to understand how predisaster information can inform a post-disaster environment. Understanding the connections and leveraging this information can be a difference maker.

How can spatial support systems help emergency managers in real time? By using data and graphically displaying how current conditions can affect disaster response and recovery, policy makers and communities can begin discussions on building community resiliency.

---

[*] The questions related to this citation are sourced from the Mills and Warren article.

Communities with low levels of jurisdictional capital are the most dramatically impacted during disaster and are the last to recover. The structural impacts of communities with low jurisdictional capital contribute directly to community resiliency.

Perhaps the best place to look to implement specific spatial decision support mechanisms is the insurance industry. The insurance industry aggregates information in a way to understand risk. Risk tolerance and management inform insurance prices. As Li, Wan, and Wang note:

> By pooling individually unacceptable risks into a far less risky portfolio, natural disaster insurance plays a significant role in disaster prevention and mitigation, lessening burdens on governments, accelerating reconstruction of disaster areas and acting as a tool for land-use planning reform by accounting for risk. (Li, Wang, and Wang, 2005)

INSURANCE INDUSTRY AND GIS TECHNOLOGY

The insurance sector must protect itself from large risks by pooling risk in a profitable manner. Government should first understand aggregate risk and then pool resources to mitigate risk in a manner to reduce capital outlays after disaster. This strategy would also increase community resiliency.

Insurance pricing is based on unit risk, aggregate risk, and hazard frequency. In a nutshell, insurance companies understand that the product they sell helps protects people from a variety of risks. But in order to be profitable, insurance companies must not give away the bank every time there is a car accident, fire, or disaster event. They must diversify risk in a manner that provides value to the customer and minimizes risks. They utilize complex and robust mathematical modeling and GIS software to assess risk and the potential for high payouts. Government should inherently understand that higher vulnerability levels will lead to higher costs post-disaster. And government also understands that prevention greatly reduces costs. Allocating or diverting resources to strengthen physical critical infrastructure is often much easier than increasing community resiliency. But if we use insurance risk methodologies, we can take a new look at social problems impacting communities prior to disaster.

Many industries have leveraged GIS technologies to map and store data points. But the GIS software often does not provide the in-depth analysis required to assess risk. Many insurance companies have leveraged GIS data with spatial support decision systems (SDSSs). SDSSs compensate for the deficiencies of traditional GISs and provide powerful and flexible facilities for integrating special knowledge (such as mathematical models and spatial autoregression) and declarative spatial knowledge (Wen, Lin, Chen, and Ming, 2007). That is, the SDSSs can take complex and fuzzy data points and help you make sense out of what you are seeing.

Governments should leverage spatial data to do the following:

- Hazard simulations across communities: Hazard simulations can help municipal authorities by understanding how various neighborhoods would respond to various disaster events. In earlier chapters you learned that there are communities within a jurisdiction. Not every community is homogenous. If we can understand that, then we should also understand that heterogeneity will lead to varied disaster preparation, mitigation, response, and recovery. But in order to understand that, we must catalogue, map, and track jurisdictional capital.
- Community-specific risk simulations: These risk simulations would test disaster simulations against specific communities. An emergency manager would be able to understand how a specific community or block would respond and recover post-disaster. This level would allow emergency managers to add or subtract various jurisdictional capital data points.
- Aggregation of community-specific risk to understand municipal and regional risk: If you can understand block-by-block risk, you will be able to understand the full scope of vulnerability across your jurisdiction.
- Exposure analysis: By leveraging your vulnerability analyses for your jurisdiction against jurisdictional capital data, you can begin to make more informed decisions across the entire emergency management cycle.

By using the following mechanisms, a local government can begin to see the intersections between municipal services, jurisdictional capital, and private sector infrastructure and resources.

SUPPLY CHAIN MANAGEMENT AND GIS

Disaster events often limit the ability of organizations to effectively get resources and aid to affected populations. GIS technology can be leveraged with social intelligence to improve supply chain efficiency. Getting resources from Point A to point B is crucial during any disaster event. But far too often, government pushes out resources without thinking about how effective those strategies are.

Natural catastrophes can cause economic disasters (Stecke and Kumar, 2009). In the United States, heat, tropical storms, floods, severe weather, blizzards, wildfires, and ice storms cause over 90% of the total economic losses from natural catastrophes (Ross, 2003). We are seeing an increase in extreme weather, and as a result, supply chains are at great risk. Population movements and the distribution of population when compared with infrastructure have increased national risks. The U.S. Census Bureau estimates that the population of coastal areas in Florida, which are highly vulnerable to hurricanes and tropical storms, has increased from 1 million in 1940 to over 10 million in 1990 (Stecke and Kumar, 2009). What we know is that infrastructure has not been developed to address this change.

GIS technologies leveraged with SDSSs can help emergency managers build more robust understanding of supply chains. For example, if there is an increase in immigration, gentrification, or other demographic movements, the resources required during and after disaster may differ from those of a more stable demographic community. Connecting jurisdictional data with existing infrastructure can provide an emergency manager with a better understanding of how to allocate resources.

This section focuses on how municipalities can use various financial mechanisms to better prepare their jurisdictions for disaster. We will discuss how local government emergency managers can work with policy makers and elected officials to craft budgets, issue debt, and deliver services in a way that increases community resiliency. Social intelligence data points can help emergency managers understand the connections between municipal services in a way to identify tipping points for more effective service delivery.

By the end of this chapter, you should understand:

- How information technology can help emergency managers better understand their communities
- How GISs and other federally developed tools can aid emergency managers in implementing SI
- How social intelligence data points (BDPs), when leveraged with GIS technology, can help you prepare more robust emergency management systems
- How existing technology tools can help emergency managers better understand the communities they serve prior to disaster

174

Again, information technology can aid emergency managers in implementing social intelligence. But first, it is important to identify the BDPs and develop relationships to obtain the information related to the BDPs. This requires buy-in from leaders in your community and building relationships within the public, nonprofit, and private sectors. Once that is complete, it is important to identify which tools can help you develop spatial and visual understanding of what is happening in your community. Software tools like GISs can help map out BDPs and ensure emergency managers can gain greater understanding of how issues across public, nonprofit, and private sectors are interconnected.

7

Social Intelligence for Emergency Management
A New Paradigm

It's time that communities implement social intelligence to create smarter emergency management systems. This book has focused on the following subjects:

1. Providing analysis on the current state of emergency management and how it must evolve to ensure communities are more resilient.
2. Providing detailed case studies on a spectrum of disasters (hurricanes, floods, earthquakes, oil spills, etc.). These disasters span nearly a century.
3. Highlighting in each case study common issues emergency managers dealt with after every disaster.
4. Arguing that the origins of American social welfare policy are rooted in disaster relief policy. Most importantly, the book has argued that disaster policy is social welfare policy, and that modern disaster policy is based on age-old perceptions of populations that may be seen as deserving and undeserving.
5. Providing a history on intelligence, a history of intelligence from a global perspective, and then details on how modern intelligence operations work. This is a parallel to the motivations of using intelligence, and how this model can be ideally suited to an emergency management context.

6. Providing a basic architecture on how to develop baseline data points (BDPs) to be used in implementing social intelligence. Chapter 5 identified BDPs every community can utilize to gain greater understanding of preexisting conditions. The chapter also provided a template to create buy-in from leadership (elected officials, city managers, department heads) to drive social intelligence implementation.

7. Providing a comprehensive understanding that geographic information systems help emergency managers create spatial and visual understanding of conditions prior to, during, and after disaster. Chapter 6 provides insight on how geographic information systems (GISs) can be used in conjunction with social intelligence.

Implementing social intelligence requires emergency managers to identify, collect, aggregate, and synthesize data. In addition to leveraging data, emergency managers must understand how these data play out in the community. In the same way intelligence agencies deploy operatives to get a better understanding of how collected intelligence reflects on conditions on the ground, emergency managers must also identify ways to connect with the community to gain a complete understanding of the communities they serve.

In Chapter 5 we discussed baseline data points (BDPs). In order to obtain some of the information, the book called for emergency managers to develop relationships with emergency operations center (EOC) partners (to obtain buy-in) prior to disaster. Developing these relationships would also help emergency managers gain access to critical information and information related to BDPs. The first part of this chapter will focus on how emergency managers can use social intelligence to build resiliency at the local level and harness preexisting community strength prior to, during, and after disaster. The first part of this final chapter will identify hyperlocal partners and strategies to fully implement social intelligence. The final half of this chapter will identify community-wide efforts to fully implement and leverage social intelligence.

HYPERLOCAL SOCIAL INTELLIGENCE

For social intelligence to work, communities must be engaged at the local level. Too much of government policy making and emergency management planning occurs centrally and without engaging the people the policies and programs serve. Previously, we outlined a social intelligence implementation strategy. To be successful within the local unit of government, buy-in is needed from elected officials and other leaders. At the local level, emergency managers must engage and partner with communities by developing relationships and partnerships with institutions and organizations that work with communities.

The first half of this final chapter provides an understanding of the types of local capacities emergency managers can harness and provides strategies for linking these capacities into the social intelligence system for a community. In turn, emergency managers will have the ability to help drive policy and programs from the ground up and top down.

Local capacities discussed in this chapter include

- Block clubs and community organizations
- Religious and faith-based organizations
- Chambers of commerce
- Local historical societies and museums
- Local post offices and banks

BLOCK CLUBS AND COMMUNITY ORGANIZATIONS

Communities and neighbors who know one another are communities that have built-in resiliency. The problem is that emergency management has not fully understood how neighbors and communities can help emergency managers build better plans. For example, while many of the stranded New Orleans residents lived in concentrated poverty, emergency managers were not able to fully harness the strong social networks in the community prior to disaster. These social networks linked generations of families and neighbors—these social bonds enabled even the poorest residents to get by when things got really rough. Perhaps understanding these bonds could have saved lives during Hurricane Katrina, and understanding these connections would have most certainly helped recovery efforts.

Many communities around the country have developed block clubs and community organizations. In most cases, the block clubs and organizations were often formed in response to a particular issue: public safety, controversial zoning, and in some cases, annual block parties. In some cases, block clubs and community organizations form to keep a watchful eye on children. Other local efforts sprout up to address environmental issues or for general beautification. The point here is to promote activities that build community and community efficacy.

Collective efficacy therefore is the linkage of cohesion and mutual trust with shared expectations for intervening in support of neighborhood social control.*

Research indicates that hyperlocal community organizations tend to promote a better general welfare within communities. A 1989 study found that aspects of neighborhood social organization resulted in mutual support for children, created more dense friendship networks within a community, and contributed to lowering criminal deviance.† In addition to community-based efficacy neighborhood and block organization activities, these local organizations have the ability to connect disconnected populations to local government and services. Senior citizens are one such example.

For many seniors, a digital divide prevents them from accessing information and services from their local government. Individuals and families without broadband connections must also deal with this reality. Block clubs and neighborhood organizations can serve as the vehicle for getting information to individuals and families. Helping neighbors meet and develop social bonds with their neighbors enables emergency managers to build more resilient communities. Emergency managers must also understand where these resilient communities already exist. That is, they must understand where communities are already working together on common causes. By harnessing this information, emergency managers can develop more informed plans and help connect with vulnerable populations in a way that is more impactful.

Emergency managers can build additional resiliency at the neighborhood level by addressing the small emergencies that often lead to major issues. For example, block club members should check up on seniors during heat waves and power outages. A well-being check during a power outage or heat wave can be a matter of life or death. Another example is to

---

* Ohmer, Mary, and Elizabeth Beck, Citizen Participation in Neighborhood Organizations in Poor Communities and Its Relationship to Neighborhood and Organizational Collective Efficacy, *Journal of Sociology and Social Welfare* 33(1): 179–202, 2006. Academic Search Complete, EBSCOhost (accessed September 22, 2013).

† Ohmer, Mary, and Elizabeth Beck, Citizen Participation in Neighborhood Organizations in Poor Communities and Its Relationship to Neighborhood and Organizational Collective Efficacy, *Journal of Sociology and Social Welfare* 33(1): 179–202, 2006. Academic Search Complete, EBSCOhost (accessed September 22, 2013).

encourage block clubs and neighborhood organizations to develop teams to help seniors and disabled neighbors shovel snow or clean catch basins. These initiatives build community and trust, and prevent a snowstorm from causing injury to a neighbor. Other initiatives include creating or promoting Neighborhood Watch programs.

Social Intelligence Implementation Strategies: Block Clubs and Neighborhood Organizations

To develop local capacity, emergency managers do not have to start from scratch. Again, many communities have built-in strengths that go unmeasured because policy makers and emergency managers do not ask the right questions or know where to look. To build local capacity, emergency managers must do the work of identifying existing capacity. This means getting out into the community and understanding what preexisting community organizations exist.

- Identify existing block clubs and neighborhood organizations through local elected officials. Using the local elected official as a conduit to their community helps emergency managers ensure buy-in from the official and ensures the initial connection. In addition, leveraging the local elected official can reinforce the emergency management office to local communities.
- Once connected with the block club or neighborhood organization, develop a community organizing kit to help socialize emergency management planning efforts.
- Where there are no block clubs and organizations, emergency managers should deploy personnel and organizing kits to help create new capacity in communities. In other cases, partner with a religious institution, YMCA, or other organization that works with the community.
- Emergency managers should work with local organizations to set up well-being check programs.
- Emergency managers should promote green initiatives by promoting the development or growth of block clubs and neighborhood organizations. For example, partnering with the local water department to install rain barrels on all homes located in a block club footprint can help mitigate localized flooding.

- Emergency managers should help neighborhood organizations and block clubs connect with seniors in their community to connect potentially disconnected community members with their neighbors and government.

RELIGIOUS AND FAITH-BASED ORGANIZATIONS

As discussed previously, many communities, individuals, and families rely heavily upon the religious and faith-based institutions in their cities and towns. In some places this is the primary location for community organizing and community events, whether organizing for disaster relief or a school supplies drive, running a summer camp for kids, or holding the local farmers' market or local fund-raising initiative. In short, religious institutions form the bedrock of many communities. Commonly, a religious or faith-based institution will come into contact with people on holidays and during major life events, but this will likely happen over a person's lifetime and throughout many generations of people and families. So, there is a keen knowledge of what is happening with and to families in an area. And, beyond just the regular events of services, community events, holidays, and life events, many religious institutions are involved in social justice initiatives and addressing challenges facing a community. For example, religious institutions will often run food pantries or soup kitchens. Therefore, the entity has firsthand knowledge of how great the need of food assistance is to a community. They are the first to know when food pantry or soup kitchen use goes up or if new visitors need to take advantage of food assistance or other available social services. Many times a religious institution will be the first to identify changing trends in a community. People often trust these institutions to help them and provide safety in some confidentiality of the need, and for this reason people and families share much of their lives and needs with the institutions and those who run them. They know people, families, needs, trends, and history of an area better than anyone.

For example, after Hurricane Katrina many people needed to prove their residency and citizenship, but for older people from the South, many had no formal documentation, just a local Bible and an inscription dating their birth. The only formal record of residency or citizenship was the Bible and the inscription from a church leader. In short, these institutions are intimately familiar with the people, a community's issues, and its history. They will also be aware of other important institutions and social

182

groups that have good information on a community. And finally, religious institutions and faith-based institutions are the entities that are depended upon by government to serve in the aftermath of a disaster. Sometimes they receive funding, and other times they raise their own funds, and sometimes it is a combination of the two, but they are always at the center of any disaster relief efforts, and that is because they are ready and able to organize, but also because they know best how to serve a community, given their centrality to the life of a community and a family.

In sum, this historical knowledge can not only help inform more culturally competent emergency management plans but can also create ties with these organizations prior to disaster to ensure emergency managers can harness community strength instead of focusing on vulnerability.

Social Intelligence Implementation Strategies: Religious and Faith-Based Organizations

Religious and faith-based institutions are the real-time eyes and ears in a community. They see issues as they arise and often work with families as they deal with those issues. Simply put, building strong relationships with religious and faith-based institutions will ensure your social intelligence system is keyed into communities.

- Identify existing soup kitchens, social service organizations, and food pantries linked with religious and faith-based organizations in the community.
- Identify religious and faith-based organizations working with and serving immigrant communities. This includes working with organizations serving the undocumented community. It is important to build relationships with these organizations to ensure the BDP associated with religious and faith-based organizations is effective. These organizations are also most likely to serve those most in need prior to disaster.
- Make sure to acknowledge that religious and faith-based organizations work with all constituencies and communities. Building effective partnerships with these organizations can help inform planning efforts for vulnerable populations.
- Religious and faith-based organizations work with families and individuals over generations. In most instances, they can identify social networks and family and community resiliencies that most often go unnoticed by emergency managers. This is true of what

most emergency managers consider vulnerable communities. By forming relationships and partnerships with religious and faith-based organizations, emergency managers can harness strengths normally unaccounted for.

CHAMBERS OF COMMERCE

Chambers of commerce and other business organizations in a community can provide emergency managers with insight into employment trends and sales trends, and can help local government officials predict how the local economy may ebb and flow. Why is this important for emergency management? For example, in single-sector reliant communities along the Gulf Coast, local chambers of commerce support businesses that in turn support and depend on a major industry. As discussed in the Gulf Coast oil spill case study, the BP *Deepwater Horizon* explosion sent millions of barrels of crude oil into the ocean. As a result of the spill, the commercial fishing industry came to a crashing halt. Many small communities along the Gulf Coast relied on commercial fishing and fish processing. Many small businesses relied on a strong commercial fishing industry to keep customers flowing in. A shock to the commercial fishing industry had major impacts on supporting sectors. This included retail, hospitality, and tourism. One event had a domino effect on the local economy. Understanding the linkages between a potential disaster and the business community can help emergency managers predict the impacts of a disaster.

Chambers and business organizations can also help emergency managers and planning officials understand where local economic fault lines lie. That is, these organizations can help map out how an economic shock (recession or downturn) may impact local employment. In addition, these organizations can help emergency managers understand issues that may arise during and after disaster. Finally, business organizations can help emergency managers better target Small Business Administration (SBA) outreach post-disaster.

Social Intelligence Implementation
Strategies: Chambers of Commerce

Chambers of commerce serve as key aggregation points for the business community. Chambers often act as advocates for the business community when local governments consider new regulations, zoning laws, etc. Emergency managers can leverage chambers of commerce to get a pulse on the business community.

- Emergency managers and other local officials should convene regular meetings to understand the health of the local economy. The ebbs and flows of the economy may impact government revenue, which can impact emergency planning.
- Emergency managers should use information supplied by chambers of commerce and business organizations to supplement private sector BDPs. Data provided by chambers can also be used to supplement local government BDPs.
- Developing working relationships with chambers and business organizations will aid emergency management's understanding of single-sector communities. Understanding the connections between employment and sector reliant communities will help emergency managers better target assistance post-disaster.

LOCAL HISTORICAL SOCIETIES AND MUSEUMS

Local historical societies and museums are a valuable resource for emergency managers. These institutions collect newspaper clippings, take oral and written histories of major events, and catalogue a community's history. This information is good social intelligence. Why? These institutions are able to provide historical information on any major event impacting a community. In the aggregate, this information can help emergency managers understand why some communities trust government more than others. In addition, these historical institutions can provide context to longstanding social issues that local officials may not have knowledge of. For example, after the Northridge earthquake, many immigrant and migrant workers were fearful of accessing services available to them because of prior interactions with government. In the aftermath of Hurricane Katrina, some residents were concerned that local officials detonated explosives to break levees and purposefully flood poor communities. This fear was not steeped in paranoia—in fact, in 1927 officials in New Orleans did open the levees via explosives and flooded poor black communities in order to save predominantly white communities. Emergency response must take historical context into account. If there is a historical link to not trusting government actions, emergency managers must take note of it. Similarly, historical institutions can also provide background on positive events in a community's history.

Social Intelligence Implementation Strategies: Local Historical Societies and Museums

Historical trauma is real. Understanding how a historical mistrust in government manifests itself is important for emergency managers. It can be the difference between a successful evacuation order and what we see in the aftermath of Hurricane Katrina. Here are some strategies to implement social intelligence from local historical societies and museums:

- Earlier in this book, we recommended developing a historical analysis of major events impacting a community. This analysis should briefly cover major economic events, disasters, and positive/negative stories of significance. This analysis is to provide a primer for how people in the community remember these events and will provide valuable detail on past community events that could help inform future planning efforts.
- Emergency managers and public works officials should develop a historical analysis of major public works projects. While schematics, bond documents, and other technical records of public works projects may exist, local historical societies and museums can provide insight into community reactions. This information can prove invaluable as a community thinks about greening its infrastructure to better handle storm water, create more green spaces, etc.
- Emergency managers and other officials should convene meetings with historical museums and organizations as they plan and conduct emergency plans, exercises, and other activities.

LOCAL POST OFFICES AND BANKS

In medium to smaller-sized cities/towns a local post office or local bank can have good information about a community. When looking to gain information in a preparedness or mitigation capacity, a local bank can tell you a lot about the local housing market, which in turn tells you a great deal about the makeup of a community and its health. For example, if a community has recently had a high number of foreclosures or defaults, this is critical information about the stability of a community and its residents. Knowing if a property is bank owned or owner occupied, or how many vacant properties reside in a neighborhood, is critical information both prior to and after an event. Also, in some areas, mail is not always delivered directly to a home, and people need to go to the post office to pick it up. It is useful to know how many remote residences are part of a city, and depend on some city infrastructure and services, but might not be easily accounted for.

Banks and post offices can also prove an important point of contact after a disaster, especially when people need to verify their addresses and their residency, and regain important documents. After Hurricane Katrina, post offices and banks were often used to verify identity and residency for people whose homes were gone, along with all their documentation. There were instances where post office or bank employees were asked to sign and notarize sworn statements of residence for older Gulf Coast residents who lost their homes and any identification. In order to receive assistance, individuals need to prove residency, so the local post office or bank, knowing the individual, was able to be a formal reference.

For many towns or smaller cities there will be one or two local post offices and banks that all residents use. These are resources that most everyone in a town will use and need on a regular basis, so they can be extremely useful. Additionally, as a federal agency, the U.S. Postal Service might be the only federal agency presence in a community, and can therefore function as a location for federal processes, such as getting forms for assistance or new identification.

Social Intelligence Implementation Strategies: Banks and Post Offices

Knowing the central nature of a post office or bank to community planning can easily be done around these entities and replicated to areas of any size.

- Emergency managers and policy makers should regularly convene regular meetings with local postmasters to understand how communities utilize postal services.
- Emergency managers and policy makers should create roundtables with local banking leaders to understand how people use bank services.

THINK LOCALLY, ACT GLOBALLY

The second half of this chapter is devoted to recommendations and strategies for implementing and utilizing social intelligence at a community level. Using social intelligence at this more global level and leveraging local SI capacity will enable emergency managers to ensure emergency management plans are holistic and inclusive of all members of a community.

Much gets said about community resiliency and how to build resilient communities. For many years, community resiliency meant hardening critical infrastructure. After disaster, government at all levels looked at strengthening levees, building seawalls, rewriting building codes, etc. These actions will certainly save lives when the next disaster strikes. It's also safe to assume that communities have learned from previous experiences with disaster, and as a result, many lives have been saved. That said, little has been done to build more resilient individuals and families.

One key element of social intelligence is using information to reduce vulnerability and to build more resilient communities. What is a resilient city?

> According to a definition by resilientcity.org, a resilient city "is one that has developed capacities to help absorb future shocks and stresses to its social, economic, and technical systems and infrastructures so as to still be able to maintain essentially the same functions, structures, systems, and identity."
>
> The term resilience is therefore looking beyond sustainability, but incorporating the values of sustainability in every aspect. Furthermore, the inclusion of the social and economic impacts points to the recent major global recession, which devastated many cities, in particular, those dependent and reliant on a single sector or industry. Building resilience can therefore look beyond the impact of climate change, and environmental disasters, and begin to factor in the social and economic resilience which can be as important.*

To summarize, emergency managers, policy makers, and elected officials need to understand their communities in advance of disaster. More importantly, they must divorce perception and policy frames from how they allocate resources and drive public policy. Implementing social intelligence can help communities take more objective views on building more resilient communities.

The second half of this chapter will focus on leveraging social intelligence to take action in your community.

---

* http://thisbigcity.net/building-resilient-cities-our-next-citytalk/.

Socially intelligent prescriptions include

- Budgets, capital planning, and disaster
- Green infrastructure
- Investing in vulnerable communities
- Investing pension funds in local projects (see NYC post-Sandy)
- Zoning policies
- Open Government—Open 311, predictive analytics, and BDPs to inform government decision making
- Creating CCOAD/VOAD organizations
- Lessons learned from recent events
- Municipal credit ratings
- Unique social factors due to location
- Public transit
- Language access and historical issues related to immigration
- Institutionalizing social intelligence

BUDGETS AND DISASTER

Until there is a federal disaster declaration and federal money flows into local coffers, communities must fend for themselves during and immediately following a disaster. Just as important is to understand that preexisting financial conditions are a reflection of preexisting social conditions in a community. How local governments manage, plan, and spend money is a reflection of their priorities. Social intelligence can help remove some historical and current policy frames that drive public policy. That is, the book earlier explained how public policy and politics are a reflection of long-held perceptions. Disasters are expensive and have cost taxpayers over $1 trillion. Utilizing social intelligence can help reduce that outlay if more thought is put into government functions and policies on the front end. This section will focus on preexisting financial conditions in a community and detail social intelligence strategies to better understand a community's financial health prior to disaster.

Many times the book has detailed the costs related to disaster. This section will once again detail why the current approach to disaster spending is not effective and provide emergency managers with budgetary and fiscal strategies that leverage social intelligence to build more resilient communities.

The cost of Hurricane Katrina was estimated at $250 billion by December 2005 and amounted to $2,027 per household (Crook, 2005). It was also in 2005 that economist Brad DeLong put the chance of a major U.S. financial crisis at 20%—the former fed chairman put the chance at 75%, and Stephen Roach (economist at Morgan Stanley) put it at 90%. What we know is that by 2008, the U.S. economy was in a freefall. This freefall was caused by real estate speculation, bank deregulation (in the repeal of Glass-Steagall), deficit spending, and the impacts of globalization.

For the emergency manager, understanding these larger global, national, and local trends is important. In the example of Wilmington, Ohio (Chapter 6), we talked about the impacts of one factory closing on a community. But another example of the toll the economic crisis has taken is the Chicago response to Hurricane Katrina. In the Chicago response case study (Chapter 2), you'll notice that the City of Chicago was able to successfully resettle nearly 10,000 evacuees from the Gulf region. Chicago was able to do so in part because the city had yet to experience the economic impacts of the downturn, and the real estate market was still performing at high levels. Chicago was experiencing annual budget deficits, but it was able to plug smaller deficits with one-time revenue schemes, or by raising revenues or drawing down reserves. In 2010, Chicago passed a budget that closed a $655 million budget deficit. It did so by drawing down funds generated from selling city assets and making minimal service cuts. If a disaster event elsewhere in the nation required Chicago to assist in resettling disaster evacuees, Chicago would not have the financial capacity to do so. The economic landscape in Chicago has changed since its 2005 Katrina response, and that same response could not be repeated today. A city like Detroit, which has gone into bankruptcy, is another example of where a community has fallen on bad times, a location that might otherwise have helped others but now is unable to help itself.

Social intelligence and greater understanding of BDPs can help emergency managers more effectively communicate to city managers and elected officials about the importance of efficient and equitable service delivery, infrastructure investments, and greater attention to preexisting social vulnerability. Disasters can strike at any time, and communities often have to deal with the cumulative impacts of vulnerability instantaneously. Governments often do not have the ability to set aside dollars beyond a fund balance reserve to prepare for disaster. Fund balances are often set aside for unexpected budget deficits and to bolster bond ratings. While setting aside funds may prove unrealistic, governments can begin to deliver services in equitable ways.

Local governments collect tax revenues from a variety of sources. They also collect revenues through fees and fines. The bulk of revenues are often generated from property taxes, sales taxes, and corporate taxes. We can further demarcate sales and corporate taxes, but for the emergency management context, we just need to understand how these funds are allocated. Generally speaking, about half of collected property tax dollars are used to service long-term debt and fund pensions and current employee costs. The other half funds basic services. In addition, corporate and sales tax revenues, along with fees and fines, are used to fund municipal services. The point is that there is just one large pot of money for a municipality. Given that there is one pot of money, one could assume that dollars are distributed equally across a jurisdiction. In theory this would be the practice, but we know in reality municipalities allocate resources unequally.

If municipal development were homogenous, there would be no need to collect social intelligence. But we know that there are often good and bad parts of a community. The good parts have well-maintained roads, parks, and schools. The bad parts often have infrastructure that is in poorer condition, schools that perform at lower levels, and property values that are often lower. Recognizing this type of unequal allocation is key, and taking action to ensure all communities receive equitable resources will ensure communities are more resilient prior to disaster.

To increase budget effectiveness, emergency managers should participate in the development of their community's annual operating budget. Emergency managers have a unique perspective in that they can see the interconnectivity between the various elements that lead to disaster vulnerability or disaster resiliency. Emergency managers should involve themselves in budget forecasting programs whenever possible.

Long-Range Forecasting

Long-range forecasting techniques focus primarily on the long-term health of the local economy—which will in turn affect the fiscal condition of local government.* Budget analysts and directors often rely on historical data or past trends to predict revenues. They may utilize economic indicators from previous fiscal years to predict revenue growth or decline. Service demand increases or decreases are also taken into

---

* *Management Policies in Local Government Finance* by Jay Richard Aronson, Eli Schwartz ICMA University, 2004—Political Science.

account. Some of the long-range forecasting mechanisms outlined by the ICMA University include

- Population projections
- Location quotient analysis
- Shift-share analysis
- Fiscal impact analysis

Population projections are important for long-range forecasting. Anticipated population is an important determinant of the future of a local economy.* The makeup of a population can have significant impacts on revenues and service demands. Communities can change rapidly, and the service needs of people vary. Communities may age—the service needs of a community today will drastically differ 20 years from now. Some communities are filled with starter homes with young families that come and go. Other anchor communities may have a core group of residents that stay for a generation or more. Some communities accept refugee populations from around the world. These populations often require acute mental health services, specialized social service provisions, and other services not required by the general populace. The examples are numerous, but even slight changes in the population size or makeup can have dramatic impacts on revenues and budgets. Understanding these shifts can help jurisdictions allocate resources.

Population projections are also useful in hiring and retaining municipal employees. Many municipalities have defined benefit pension programs. Nationwide, pension systems have been grossly underfunded. There is an estimated $1 trillion in unfunded pensions nationwide. Pension payments will stress already tenuous budgets—many municipalities will be forced to make drastic service cuts. The demand for service, however, will not diminish. So understanding the population (and underlying demographics) and any shift is paramount going forward. Emergency managers should carefully monitor population projections relative to the baseline data points.

Population projections can inform the baseline data points in the social intelligence scheme we outlined earlier. The data could flow both ways—data could be plugged in to the various baseline indicators or simply reported upstream to policy makers and elected officials. Similarly, data reflective of the population projections could be disseminated

---

* *Management Policies in Local Government Finance* by Jay Richard Aronson, Eli Schwartz ICMA University, 2004—Political Science.

downstream or cross-stream to municipal, nonprofit, or corporate partners. Examples of how population projections and social intelligence indicators can provide budget directors with more information follow:

- Police: Population projections will have an impact on police hiring, retirements, and pension obligations.
- Fire: Fire coverage will vary by population projections and density.
- Public works: Infrastructure will require varied levels of improvements based on upward or downward population shifts.
- Administration: Elected officials and administration officials can track real-time changes in tax/revenue collections to inform budgets of more robust resource allocations.
- Building: Population projections can inform better zoning and development. Development will help inform public works of capital improvement planning and can also inform school systems of potential increases/decreases in students. Building can also inform public safety (police and fire) of potential hiring, diversity training, special needs equipment purchases, etc.
- Finance: A finance department that is better informed on population projection can then accurately predict fiscal health relative to homeownership, sales tax revenues, corporate revenues, etc.
- Education: School administrators can plan for upticks or reductions in students. They can also identify emerging requirements for various populations, such as ESL programs, etc.
- Hospitals: Not many municipalities own/operate healthcare systems, but some still do. Population projections can help administrators prepare culturally competent information on illnesses, prevention programs, etc.

These are examples of downstream intelligence dissemination. The intelligence from the baseline indicators can also flow upstream. For example, public health surveillance professionals can report upstream to municipal managers about upticks in specific illnesses that are often isolated within socially vulnerable communities. Tuberculosis or staph infections (MRSA) largely impact poor communities in urban areas. Quite often the populations in these areas have limited primary care access or inadequate health coverage. A small infection or illness can metastasize into something much more rapidly in poorer neighborhoods and spark a public health emergency. A doctor, nurse, or public health professional may notice an uptick in a specific type of medical treatment and then report the trend via the appropriate social intelligence indicators.

The municipal manager can then disseminate information to the proper authorities. The budget director may then be able to better fund prevention efforts and programming, and allocate resources. In the long term, the budget director can work with other governmental and nongovernmental partners to identify the structural causation of the underlying vulnerabilities that triggered the illness.

Location Quotient Analysis

Location quotient analysis (LQA) is a popular analytical technique used to obtain an understanding of the local economy's industrial composition.* The LQA studies the overall importance of an industry or set of industries within a jurisdiction relative to their upstream and downstream partners. How is this important for emergency management? In Chapter 6, Wilmington, Ohio, was discussed as a company town. Single-sector reliance can help communities thrive, but can also serve as an Achilles' heel.

Economic development should be holistic, though in practice economic development is ad hoc. Long- and short-term plans are developed, but elected officials and policy makers have not fully appreciated the need for industrial and commercial diversity. Economic booms and busts have direct impacts on how you, as an emergency manager, plan for disasters. Booms and busts have dramatic impacts on government revenues. These revenues fund your programs. Social intelligence can help you better inform your upstream and downstream partners of emerging trends in employment, unemployment, demographic shifts, etc.

Case Western Reserve University economist William T. Bogart argues that a useful way of viewing a metropolitan area is as a collection of small open economies that specialize and trade with each other. (Bogart and Ferry, 1999). This definition applies to all municipalities. Metropolitan areas, suburbs, and rural communities all connect into an economic ecosystem. These ecosystems are in a state of homeostasis. The local homeostasis is based on a delicate system of interconnected parts that exist in symbiotic harmony. A slight disruption in one of the parts can have a dramatic, even devastating, impact on the whole system.

A good example is the City of Chicago. For decades the city was able to piecemeal services without paying attention to service structure or sustainability. With the collapse of the real estate markets in 2008, Chicago

---

* *Management Policies in Local Government Finance* by Jay Richard Aronson, Eli Schwartz ICMA University, 2004—Political Science.

faced an uncertain future. The city had been overspending for nearly a decade but was always able to plug deficits with one-time revenue schemes, increases in fees, or taxes. By 2010, the city was forced to sell public assets to cover budget deficits. This was a fiscally poor decision, and elected officials deferred making tough decisions on service structure. A disruption in the real estate market has had a profound impact on school districts, public works capital programs, and staffing. City employees now take 30 days of furloughs. The Office of Emergency Management and Communications (OEMC) has seen its budget cut and staff reduced. Police manning is short 2,000 officers daily. So a real estate crisis has had a tectonic impact on the city's fiscal health. Property tax collections are down as foreclosure rates have spiked. The resulting impact on property values has caused a series of strategic defaults among homeowners. The real estate transfer tax revenue stream has suffered due to sluggish sales. National trends have also hurt Chicago. Chicago had been a favorite destination for conventions, but also one of the most expensive. High labor costs scared many conventions away, and as a result, millions of dollars in tourism revenues were diverted away from Chicago to more cost-friendly cities.

Conversely, there can be positive shocks within a system. High-tech and biotech firms have begun to relocate to Pittsburgh. Until recently, Pittsburgh was struggling. Steel factories left decades ago, and the industry leaving had dramatic impacts on the city's financial and social systems. But recently, companies attracted to Pittsburgh's transportation network and strategic location sparked an economic surge. The result: the Pittsburgh economy is now thriving and the city boasts a thriving downtown. City revenues have climbed, and people have started to move back in to the city. LQA should be utilized to inform elected officials and policy makers of real-time trends in your community and surrounding communities. Emergency managers and policy makers should survey their community for trends in the private sector as well as neighboring communities. To do this, emergency managers will not require specialized skills, analytic tools, or computer systems to analyze communities. The social intelligence indicators are often just data extracts from much larger reports that are published by upstream/downstream partners. Emergency managers have to extract that information and connect the dots. By connecting the dots, emergency managers can work with budget directors and policy makers to make more informed analyses decisions.

Shift-Share Analysis

Now it is possible to see how revenue forecasts are tied to local, state, national, and even international trends. While location quotients provide information about a local economic base for a given point in time, a shift-share analysis attempts to describe how, over time, local employment has changed relative to national employment.[*]

A shift-share analysis considers changes from three different perspectives:

1. How much local employment would have grown had it increased relative to the national rate
2. The growth or decline in local employment in a given industry, given that industry's relative growth or decline nationally
3. The degree to which local employment within an industry grew as rapidly as employment within that industry nationwide (Aronson and Schwartz, 2004)[†]

From an emergency management standpoint, this information is vital in recognizing emerging trends. These trends can inform emergency managers of emerging vulnerabilities or identify opportunities for increased attention to infrastructure or social resiliency. National employment trends are often reflected by industry. While an impacted industry may not impact your jurisdiction or municipality, as an emergency manager you should know how regional employment issues impact you.

Municipalities often incentivize growth through various financial mechanisms. They might leverage tax incrementing financing, tax breaks, or industrial revenue bonds to jump-start industrial or corporate sectors. These tax breaks have positive impacts but can also have unintended consequences on other municipal services. One common impact of economic growth is the service requirements of high daytime/low nighttime populations in a municipality. The economic growth was spurred by a tax incentive—the increased daytime population increases risk and demand for public safety coverage. If the economic hub of the community is centrally located, it is possible that workers may live outside your jurisdiction. You may be faced with unequal levels of demand for service, and that might stress your emergency management budget. Planning efforts

---

[*] *Management Policies in Local Government Finance* by Jay Richard Aronson, Eli Schwartz ICMA University, 2004—Political Science.
[†] The three perspectives for shift-share analysis are all sourced from the Aronson and Schwarz ICMA textbook.

will certainly vary by daytime/nighttime standards. These are issues that emergency managers already grapple with. That said, emergency managers should also begin to harness social intelligence indicator data points as they relate to local, state, and national employment trends to ensure living plans are updated in real time.

Emergency managers are increasingly dealing with disasters that require long-term recovery. Employment trends are becoming more important in structuring recovery efforts. Emergency managers must begin to understand how local education systems, social networks, and other structural factors contribute to jurisdictional capital. If an emergency manager has a predisaster understanding of the local employment markets, it may be easier to relocate people post-disaster according to skill sets. The more you know about the structural factors of employment in your jurisdiction, the better off you will be post-disaster.

Fiscal Impact Analysis

Fiscal impact analysis (FIA) is utilized by municipalities to leverage shift-share, LQA, and population projections. The FIA attempts to connect the dots from a fiscal perspective. The FIA is used to determine infrastructure, police, fire protection, and other service needs by specific segments within a community.

By way of example, the State of Utah developed an input-output-based economic model that allows forecasters to estimate how changes in employment and output in one industry impact other industries (Aronson and Schwartz, 2004). This same FIA model should be developed for municipal services. Changes in inputs and outputs by increases/decreases in revenue streams will have dramatic impacts on service delivery and staffing. It is very necessary to understand how governmental revenue streams will impact industries, but those revenues also impact governmental services.

Emergency managers can leverage the social intelligence indicators to compare revenue data points against specific services. For example, emergency managers can determine what a funding increase or decrease can do to a capital improvement program within the police or fire department. Emergency managers can also understand how an overall increase in economic output can lead to increased revenues. An increase in economic output may spur populations to growth in parts of their jurisdiction. Service demands for a more densely populated area may exceed current capacity. The revenues generated from that surge may not be adequate relative to service demands. This is where emergency managers can have preemptive discussions about community growth or contraction.

Budget directors, administrators, and elected officials view municipal finances from 10,000 feet and then get down into the weeds. Emergency managers are often focused on the problems on the ground—they see themselves as first responders and often focus on issues in the weeds. While that approach has worked, emergency managers are uniquely qualified to join the 10,000-foot view discussion.

Capital Improvement Planning

Capital improvement planning is a local government's way of planning for public works improvements, equipment purchases/upgrades, maintenance of infrastructure (roads and bridges) or equipment, and purchase of new assets. These expenditures are usually large and are meant to benefit the public over a considerable period of time.*

Capital plans are built by examining the revenue and expense sides of the budget. Revenue streams are examined relative to economic growth or contraction and then compared to existing plans for improvement. Capital budgeting rests on a foundation of economic base studies, land use reports and maps, and population and migration studies, which provide the basis for projecting fiscal resources, assessing need, and identifying the most productive and viable capital projects.

Policy makers should then be able to leverage social intelligence and GISs to add value to current capital improvement plans.

Specific questions to ask as part of this process include

1. Are there specific infrastructure degradation patterns over time that contribute to social vulnerability?
2. Are there connections between infrastructure and property tax collection?
3. Is infrastructure to support current industries adequate?
4. Are schools, parks, and other public service facilities adequate?

Capital improvement plans help policy makers make decisions about maintaining and repairing infrastructure.

---

* *Management Policies in Local Government Finance* by Jay Richard Aronson, Eli Schwartz ICMA University, 2004—Political Science.

Debt and Local Government

Local governments often issue municipal bonds or long-term debt to finance large projects. They intend to spread out the cost of the project over the life of the end product. Normally tax revenues or revenues generated from the project are used to pay down the principal and interest. The fiscal health of a local government is reflected by the bond rating issued by Moody's, Fitch Ratings, and Standard & Poors. These credit ratings impact interest rates, repayment terms, and insurance coverage. The better the financial health of a local government, the better the credit rating.

By now you understand that a number of social intelligence indicators factor into fiscal health. The health of a municipality can fluctuate, as with the health of the economy. Revenue streams ebb and flow with the economy. If you are able to understand these ebbs and flow in real time, you can use this information to take advantage of bond markets. Refinancing debt and issuing new debt is a way local governments realize cost savings. By utilizing social intelligence, you may be able to identify instances to refinance debt, extend out payment terms, or pay down debt.

GREEN INFRASTRUCTURE

As flooding, increased rainfalls, and other climate change issues impact communities to a far greater degree with far more frequency, it is important to identify ways to make infrastructure reflect a changing climate. Cities around the world are identifying ways to invest in infrastructure that ensures water supplies remain clean, addresses heat island impacts, and better manages storm water. It is also important to target the infrastructure investments in communities where social vulnerability and disaster vulnerability align. That is, policy makers and emergency managers should work together to identify areas in a community where disaster impacts are exacerbated by a historical pattern of disinvestment or a lack of investment. Second, emergency managers should map out where disaster impacts are most likely to impact communities. And third, emergency managers should work with policy makers, department heads, and elected officials to target infrastructure investment in these areas. Examples of green infrastructure investments include green alleys, green roofs, bioswales, rain gardens, permeable surfaces in streets, geothermal heating and cooling under street surfaces, and the creation of more green spaces.

In addition, local government should partner with community organizations to ensure individual efforts are championed. Some efforts include installing rain barrels, diverting downspouts from sewer systems, utilizing green roofs, weatherizing windows and doors, using green lighting fixtures, and reducing energy consumption. By coordinating investments at a municipal level with outreach and investments at the hyperlocal level, emergency managers can build more sustainable communities and ones that are more resilient to the impacts of disaster.

Using social intelligence BDPs, emergency managers should identify areas in communities for green infrastructure improvements. Some examples include

- Leverage hospital BDPs to identify areas in the community where asthma rates and heart disease rates are prevalent. These areas are ripe for implementing green infrastructure impacting air quality and mitigating air pollution.
- Leverage public works BDPs to identify areas in a community where infrastructure is aging. For example, in areas where water and sewer mains/drains are aging, target green infrastructure. This can reduce overtime expenditures if the infrastructure fails. More importantly, understanding the age, condition, and capacity of infrastructure can help emergency managers advocate for comprehensive capital plans. Even if unfunded, long-term capital plans empower elected officials and policy makers with data to advocate to state and federal officials when capital funds are allocated.
- Leverage building department BDPs to implement new zoning and building codes. One easy way to reduce the impacts of flooding is to require actual green space in new developments. Another change in the codes and regulations is to forbid builders from building structures lot line to lot line. Requiring green space and setbacks from neighboring structures ensures water can filter into the ground instead of overwhelming sewer systems.

INVESTING IN VULNERABLE COMMUNITIES

Communities with high levels of poverty are also usually located where disaster impacts are the greatest. In the case studies, Chapter 2 described how communities in New Orleans and the Gulf Coast were struggling prior to Hurricane Katrina and the *Deepwater Horizon* oil spill. Preexisting conditions drive what happens during and after disaster. Those same preexisting conditions should drive mitigation activities and recovery activities. In an ideal world, local governments would invest in communities on the front end—in most instances, government response is reactionary. In order for social intelligence to drive recovery and mitigation activities, it is important for emergency managers to use BDPs to understand the fabric of vulnerability. Some examples of increased or targeted investment include

- Leverage public transit BDPs and housing BDPs to understand where affordable housing is located in a community and map out proximity to public or mass transit. For example, in Chicago, the Center for Neighborhood Technology has developed a true affordability index. This index ties the location of affordable housing to its proximity to mass transit. Why? If affordable housing is located in far-reaching corners of a community, the rent or purchase price may be affordable, but transportation expenses are much higher. Emergency managers know that affordable housing during any disaster recovery is scarce because it is in limited supply predisaster. Emergency managers should consider transportation costs when identifying new affordable housing developments pre- and post-disaster.
- Leverage school and education BDPs to drive investment in public schools. Even very poor communities, public education, and neighborhood schools are institutions that bring community together. Emergency managers can develop a greater understanding of vulnerable communities by studying how schools perform. School performance is a useful indicator of what is happening in a given community.
- Leverage nonprofit and social services BDPs to help drive more inclusive public policy. For example, the City of London recently discovered that tying living wages to realistic expense assumptions results in better mental health for communities.* It is difficult for people to work full-time jobs and still not provide for their families. Understanding that the working poor work full-time helps deconstruct many political and incorrect frames policy

---

* http://www.chicagotribune.com/health/sns-rt-us-living-wage-20131009,0,101153.story.

makers utilize when thinking about the vulnerable. Developing public policy and living wage policies can help create healthier and more resilient communities prior to disaster.

INVESTING PENSION FUNDS LOCALLY

Most public sector jobs have pension plans. Police, fire, teacher, and labor pensions are common around the country. Employees pay a portion of their salary into the pension funds, with taxpayers paying a fixed percentage. While pension funds are not generating the needed returns to sustain themselves, there are billions of dollars invested nationally and internationally on various municipal projects, bonds, stocks, and other investment vehicles. In most cases, and to spread out investment risk, pension funds from one community are invested in other communities. Examples include purchasing municipal bonds for sewer projects, water treatment facilities, and road construction projects. The communities then pay a fixed rate of return to the pension funds. These investments provide pension funds with a stable return on investment in relatively low-risk projects.

After Hurricane Sandy, New York City officials moved to invest pension funds locally into affordable housing developments. This innovative funding strategy paved the way for creating homegrown stimulus projects. Most major cities have billions of dollars across many pension funds. To satisfy risk management requirements, most of these funds are invested across the country and world to reduce investment risk. That said, communities should follow the New York City model and invest a portion of local pension funds into affordable housing projects, roads, storm water management, and other projects to help develop communities. Investing these dollars locally can create many jobs, develop underserved communities, and increase the resiliency of communities prior to the next disaster.

ZONING POLICIES

Like building codes, zoning policies and ordinances play a significant role in how communities prepare for, respond to, and recover from disasters. For emergency managers, it is critical to understand how a changing climate impacts existing structures. It is just as important to capture lessons learned from disaster events and build those lessons into the zoning code.

For example, emergency managers should work with building department officials to ensure adequate open space and setbacks are enforced during new development. Other examples include

- Identify ways to link transit-oriented development and affordable housing. Access to transit and affordability in housing are linked. Creating greater accessibility to transit while building additional affordable housing stock will reduce social vulnerability.
- Emergency managers should work with policy makers to ensure new developments have adequate green space. Green space should be defined as grass or natural surfaces that allow water to filter into the ground.
- Local governments should incentivize green development or green building characteristics: green roofs, solar panels, LEED certification, etc.
- BDPs should adapt to account for green infrastructure and zoning policies.

OPEN GOVERNMENT—OPEN 311, PREDICTIVE ANALYTICS, AND BDPS TO INFORM GOVERNMENT DECISION MAKING

Social intelligence is data and information driven. The more information policy makers and emergency managers have prior to making emergency plans, the better the emergency plans. Social intelligence calls for the collection, aggregation, and analysis of information that already exists in communities. One source of information that can help inform more robust planning efforts is constituent service requests and other publicly generated data points. How people interact with their government is telling in what is happening in a community. For example, one community may consistently request water service due to low water pressure. Another community may call its local government due to crime or other illicit activities. Yet another community may interact with government on road conditions. There are obvious patterns that can be uncovered—Open Gov is a transparency initiative to unlock government data so programmers and application developers can unlock patterns and help make government more efficient while simultaneously encouraging more civic engagement.

Perhaps the best way to engage communities and use data to drive decisions is through Open 311. Many larger municipalities utilize a 311 telephone number to take service requests. Constituents call 311 and report problems and issues. These requests are then filtered to the appropriate department and responded to. Open 311 is a framework that enables people to take that service request information and do the following:

- Show taxpayers how their dollars are put to work by making service delivery more transparent.
- Provide context to service delivery. All service requests are not equal. Some issues can be addressed immediately, while others are more complicated.
- Allow people to interact on service request issues. That is, they can drop their service request on a map and see other similar reports in the area.
- Allow people and government officials to identify emerging trends.

Creating more transparent service delivery systems can help people understand how government functions. Transparency builds public trust. Open 311 is a framework to build that trust and create a useful tool for identifying emerging trends and creating efficiencies in government. For example, residents on a particular block may call 311 regarding a low water pressure. Everyone on a given block may call about the same issue. Clearly, there is a trend or an underlying issue. But depending on when the calls are made and the person taking calls, this trend may go unnoticed. With Open 311, applications can help provide spatial and visual understanding of a service request. Applications can also alert service responders to emerging trends. The low water pressure can be the result of a small leak or a large underlying failure. A large failure can spark a massive main to fail—this would result in overtime costs, unexpected expenditures, etc. Open 311 applications can help policy makers and emergency managers understand what is happening in a community. This includes impacts to infrastructure, how capital improvements impact services, etc.

Taking this one step further, predictive analytics can help policy makers and emergency managers forecast trends. The predictive technology is still new and untested, but in the near future technology will help build smarter cities.

- Emergency managers should work with application developers and policy makers to develop more robust service request tools.
- Social intelligence will be more effective if Open 311 and other mapping technologies are utilized. Utilizing these technologies

will provide spatial understanding of vulnerability, poverty, and other issues impacting communities during and after disaster.

- Capital planning efforts, capital investments, and allocation of resources are more transparent when constituents can provide input and affect the output. This creates more trust in government.

COAD/VOAD

COAD stands for community organization active in disasters. VOAD stands for voluntary organization active in disasters. The purpose of a COAD or VOAD is to build more resilient communities through the 4Cs: cooperation, communication, coordination, collaboration. In most cities and towns where COADs or VOADs have been established, the terms are interchangeable. COADs and VOADs can be very different in structure, but most generally are a coalition of nonprofit organizations that share knowledge, resources, and information regarding disaster events. The organizations often meet in all phases of the disaster cycle (mitigation, preparedness, response, and recovery) to prepare their communities for disaster and serve their communities during and after disasters. While nonprofit agencies that are active in disasters comprise the membership of COADs and VOADs, governmental agencies and departments are close partners of local COADs and VOADs. Businesses and corporations also serve as important partners to COADs and VOADs. There is a national VOAD organization (NVOAD) that helps support and develop local VOADs (www.nvoad.org).

Large nonprofits include

- American Red Cross
- Salvation Army
- Catholic Charities
- Jewish Federation
- Church World Service
- Amateur Radio Service
- Feeding America, the Humane Society
- Lutheran Disaster Response
- Nechama
- Southern Baptist Convention
- United Way
- United Methodists Committee on Relief

- Islamic Relief USA
- Local or chapter partners and smaller local nonprofits

All of these are frequently responsible for contracting/working with government to do mass care, shelter, feeding, rebuilding, and long-term recovery work during and after disasters. In order for these operations to work well, government and nongovernmental organizations (NGOs) need to know each other (i.e., have preestablished relationships). This requires knowing every organization's capacities, their areas of expertise, their reach into communities, and how they will partner. During a disaster is not the time to get to know your partners. Far too often disasters will strike a community, and government will assume there are social services available to plug disaster survivors into; unfortunately, this is not always the case. If government is going to depend upon nonprofits and religious institutions, it had better be aware of just what is available. And, NGOs that respond to disaster need to know what government will depend on them for, and what will and will not be funded. If there is not a local capacity to do certain types of disaster services, this is information that must be known ahead of time. Government cannot just hand off all the recovery work to a community that is not resourced to sufficiently help people recover. COADs and VOADs are essential community tools for government and NGOs. They build relationships prior to disaster; they convene all resources in a community, and identify areas of expertise and development. Communities that have established COADs and VOADS have proven time and time again to be more resilient communities, ones that are best prepared to save lives and property and recover as quickly as possible.

If there is a COAD or VOAD already established, join the group. If one has not yet been started, it can be convened by an NGO, a religious institution, a government entity, or any person willing to convene and develop a critical disaster resource.

LESSONS LEARNED FROM RECENT EVENTS

Whether an emergency manager is new to a community or has been living in a town/city for his or her entire life, becoming familiar with disaster events that have impacted that community is essential, and this means having information about the event, its causes, and how to make sure any mistakes made in preparedness, response, and recovery are not repeated. In short, this means becoming familiar with an event, which requires having a good background on the lessons learned post-event. In many cases, catastrophic events are common knowledge. For example, most everyone will be aware of recent events, such as Super-Storm Sandy, Hurricane Katrina, and the Gulf Coast oil spill. While these catastrophic events received significant amounts of media coverage in the days leading up to, during, and immediately after them, there is often little knowledge or coverage of important lessons learned—lessons that often cannot be uncovered for weeks, months, and in some cases, up to a year after an event.

Government organizations like the U.S. Government Accountability Office (GAO), the local office of emergency management or the mayor's office, and local NGOs will compile reports listing the many lessons learned post-event. Academic and scientific studies will often be published sometime after an event and are critical pieces of information in understanding a community, its challenges, its strengths, and how it can become better prepared and more resilient for the future. Of course, it's always important to seek out different types of reports, as political administrations can have different viewpoints and incentives in their reporting of lessons learned.

Additionally, while catastrophic events might be common knowledge, there are often smaller-scale events that might be devastating for one part of a community but completely unknown in other areas. For example, large cities or counties might have a specific area along a river that has had extreme flooding, but if it only receives minor media coverage and only happens in smaller areas of a big city or county, it does not become common knowledge. Many big cities will rely on their offices of emergency management for this historical knowledge, but when there is considerable staff turnover, layoffs, and regular changes in administration for government agencies, just a few years after an event, there might be no people left with historical knowledge about smaller or geographically isolated events. For this reason, it is critical to seek out lessons learned, find published academic and scientific studies for an area, and publish lessons learned for future emergency managers.

One example where lessons learned provide critical information for emergency management and local governments is the large-scale evacuation post Hurricane Katrina. Most people will recall that after the storm, hundreds of thousands of people evacuated to cities across the United States. But most will not know if this type of evacuation was successful, what it meant for the communities that took evacuees in, and if this type of plan should be repeated. It is not entirely intuitive that a storm that took

207

place on the Gulf Coast would provide critical lessons learned for a city 1,000 miles north.

In Chicago, we now know that we have the capability to do a large-scale recovery and resettlement program. The City of Chicago and local NGOs have a good level of preparedness to replicate this type of operation. However, we also know that resettling 10,000 evacuees is very different from resettling hundreds of thousands of evacuees, and there is little chance of this exact same event happening again. And we know that an event like an earthquake along the New Madrid Fault in southern Illinois could result in hundreds of thousands of evacuees coming to Chicago, and lessons learned from the evacuation to Chicago post-Katrina can help us adapt a very successful operation for a larger-scale yet very similar operation. Fortunately, the successes and challenges of this operation were documented by the NGOs that worked on the massive effort. This is important for gaining the necessary information, and as many of the individuals who worked on the project are no longer in their same positions and the local government has gone through major transitions, historical knowledge would not be easily accessed through people or organizations/ departments.

MUNICIPAL CREDIT RATINGS

Like individuals, municipalities also earn credit ratings. These ratings are issued by Fitch Ratings, Moody's, and Standard & Poors. These rating agencies assign credit ratings based on a community's revenues, expenditures, pension obligations (funded vs. unfunded), other post-employment benefits, outstanding debt, and general outlook. In sum, the ratings agencies survey the economic landscape around a community and then study financial indicators to inform the rating agency decision on creditworthiness. Municipalities issue long-term debt in the form of bonds to finance projects like road construction, sewers, water mains, water treatment plants, police stations, etc. The debt is paid with general revenues a community collects—property taxes, sales taxes, fees, and fines. How do municipal credit ratings impact emergency management?

Take the *Deepwater Horizon* case study. A massive oil spill off the Gulf Coast shut down the fishing, oil and exploration, and tourism industries. These industries employ people. People pay property, income, and sales taxes. If people are out of work, those tax revenues are quickly diminished. These revenues pay for municipal services and service long-term

208

debt. A shrinking revenue base could spell financial disaster for a disaster-impacted community. Emergency managers should be aware of their community's financial condition. A drop in bond rating due to rising pension obligations will make it harder for a community to borrow money. This leads to delaying capital improvements to critical infrastructure, which leads to increased disaster vulnerability.

As a community implements social intelligence, emergency managers should utilize the following strategies to maintain situational awareness of a community's financial condition:

- Study finance, administration, and corporate BDPs on an annual basis to develop a baseline understanding of the community's financial health.
- Once the initial study is complete, help inform budgetary decisions related to capital planning.
- Monitor corporate BDPs for any major shocks in the local economy. A major layoff or economic downturn would result in decreased tax revenues. This could significantly impact public safety operations and planning.
- Regularly convene meetings with local leaders, nonprofits, business leaders, and policy makers to ensure they have real-time awareness of what is happening in their community.

UNIQUE SOCIAL FACTORS AS A RESULT OF LOCATION OR SIZE

It's important to be aware of unique social factors that affect an entire community or parts of a town or city as a result of location. For example, border towns, especially those along the southern border with Mexico, can have strained relationships with government. There can be difficult relationships between levels of government and law enforcement agencies regarding jurisdictions and what laws are being enforced. There can also be a high level of distrust between those who live and work in border towns and the government entities in the area.

Additionally, there are thousands of people who move back and forth between the United States and Mexico for work, commerce, and tourism on any given day. Understanding that this is an added layer of complexity to consider will be critical in providing safety and security, and understanding the diversity of needs for border towns and their constantly changing populations.

And finally, there can be a significant number of people on the U.S. side of the border that are undocumented—this is also true for many big cities. When a disaster hits, it is important to consider that there can be a large number of people who might require help and services but cannot provide residency documentation. It is also important to consider that many people who desperately need help might not ask for it for fear of arrest or deportation. All plans need to take into account people who are disenfranchised, marginalized, living below the radar, and in the United States without documentation. Plans must incorporate strategies for doing outreach to those in need who are distrustful or fearful of government.

Other unique social factors based on location or size can include examples such as severely isolated sections of metropolitan areas. Many major metropolitan areas have seen their populations decrease over the last 10 years. This has left sections of cities isolated and, in some instances, completely cutoff from municipal services. In extreme instances there is Detroit, where entire neighborhoods have been cut off from municipal services such as water, electricity, public transportation, fire hydrant access, etc. In places like Chicago, neighborhoods have not been formally cut off from services, but people have left certain areas of the city in large numbers, resulting in neighborhoods with fewer people, more open space, vacant buildings, few businesses, and thus the opportunity for gangs to thrive. Additionally, medium and large cities have been forced to close schools in recent years. This has left some neighborhoods without any government presence at all, further isolating communities.

When looking at towns and cities in more extreme settings, such as along an international border, or big cities that have been faced with declining populations, it is important to learn about different neighborhoods within and how some are less equipped to work with government. Trust is also a major factor to consider. Neighborhoods where relationships with law enforcement are strained are areas that might be less likely to trust government interventions.

PUBLIC TRANSPORTATION

Access to public transportation is often the tipping point in economic development. In large urban areas, access to mass transit is often correlated with higher property values. In New York City, parking comes at a premium and traffic is a nightmare. Car ownership is expensive, cumbersome, and often unnecessary. The carbon emissions resulting from traffic emissions drive asthma rates in many communities. Traffic congestion is also a leading cause in hypertension—more traffic equals more stress—and many cities are searching for answers in their quest to reduce congestion. The only viable solution is to invest in mass transit and make strategic investments in communities that lack access. If access to mass transit results in higher property values, strategic investments in creating new access to public or mass transit can serve as a tipping point to reducing vulnerability in a community.

Emergency managers and policy makers should:

- Identify neighborhoods with robust access to mass transit and identify neighborhoods where transit deserts exist. This mapping exercise can be informed by the transportation BDPs. The social intelligence generated here can be used to target state and federal dollars available to mass transit agencies.
- Work with public transit officials and affordable housing experts (within government and outside government) to identify strategies to create more affordability in a community.
- Work with public transit officials to identify transit deserts. Where deserts exist, emergency managers should plan to provide emergency transit during response and recovery phases.
- Work with transportation officials and major employers to understand where employees live and how they commute to work.

LANGUAGE ACCESS, CULTURAL COMPETENCE, AND HISTORICAL ISSUES RELATED TO IMMIGRATION

Access to disaster information pre- and post-event must be communicated in as many languages as possible. While many cities have made great strides in offering information and services in popular languages such as Spanish and varieties of Chinese languages, there are still severe weaknesses in providing emergency information in the languages spoken throughout the United States. In any given year, approximately 13–15% of Americans are foreign born, and many other nonnative English speakers are in the United States as migrant workers, tourists, and students, and some are here without documentation. Additionally, languages like German, Italian, and Polish, which once dominated the foreign languages

211

most spoken in the United States, have taken a backseat to other languages, such as Spanish, varieties of Chinese, Tagalog, and Vietnamese. Of course this will vary depending on community and will continue to change as different immigrant groups and refugees seek new homes in the United States. It is imperative for emergency management to stay at the forefront of language access issues by hiring people with various language capacities and different sets of cultural competence, and who are doing preparedness work in all languages prominent in a community.

This country is, and has always been, a melting pot of people from different countries, cultures, and background. As such, we should encourage and foster this and, as part of social intelligence, account for specific populations that we know to be in our community.

Being aware of local language needs will also help in keeping up to date on immigrant and refugee populations as they settle in a community. Throughout the history of the United States different immigrant and refugee groups have chosen different cities or areas of the country to build new homes. While this will certainly ebb and flow over time, knowing that various ethnicities reside in a community will help in building vital cultural competence, which will prove absolutely necessary in determining how to communicate with and assist communities with different cultural norms, language, and backgrounds.

And finally, understanding historical issues related to immigration is important intelligence about a community. Any historical issues related to interactions with government or mistrust of government can help inform current planning, or at least provide critical insight into how planning, response, and recovery efforts may be received. For example, after the BP oil spill in the Gulf Coast, no recovery efforts took into account the very large Vietnamese population. Thousands of Vietnamese families resettled on the Gulf Coast after the Vietnam War. This was a good location for the Vietnamese, as they could restart their lives in one of the trades they knew very well, shrimping. Unfortunately, while the Vietnamese comprised a significant portion of the population along the Gulf Coast and a huge portion of those whose profession was devastated by the spill, BP, in collaboration with the government, didn't provide any information about recovery and recovery funds in Vietnamese until the programs had almost ended. This caused a great hardship on the Vietnamese families and severely stunted the ability for many to recover at all, especially for shrimpers who were still taking steps to recovery post Hurricanes Katrina and Rita.

- City council passed English as the official language. If you don't speak English, you can't access services. Fillmore.
- In San Francisco, Asians were not allowed to own land. Natural-born citizens were allowed to get disaster relief. All records were destroyed when city hall was destroyed.

CONCLUSION

Social intelligence is not a one-size-fits-all solution to developing more resilient communities. Instead, it is a culture and a way of thinking about issues. This book argues that politics and public policy are a reflection of what people think, existing social frames, and history. Using existing frames to drive emergency management plans is why emergency management continues to fail during disaster. Intelligence analysts detach themselves from the frame but acknowledge the frames in their reports. The goal is to develop fact-driven reports and analyses of what is happening in a given country. This same analytical approach can be leveraged in implementing social intelligence.

The baseline data points (BDPs) and other prescriptions are just a baseline for emergency managers. Obviously, every community is different. The people, industry, geography, culture, and history vary from community to community. All of these factors should be understood when preparing for disaster. Understanding these factors will help ensure emergency management plans reflect the nuance in a community. Social intelligence cannot be rolled out by a consultant or over the course of a few months. In order for social intelligence to work, emergency managers and policy makers must first buy in to the philosophy and framework. Next, they must understand that communities ebb and flow over time. Communities shrink, expand, and sometimes stay the same. People move in and out. Immigration policies create inflows of people. Simply put, communities are living organisms. In the same way public health officials survey a community and intelligence agencies analyze a community, emergency managers must do the same.

Emergency management must evolve. The field can no longer focus on just interoperability, command and control systems, or new gadgets. The next generation of emergency managers must focus on people and the systems they interact with. As emergency management evolves, so too must the academic research that surrounds disasters. Emergency managers should leverage the sociological inquiry related to disasters as they build emergency management systems. Social vulnerability to disaster must be understood. Social intelligence is a methodology and framework that can help the next generation of emergency managers make sense of how preexisting conditions will impact communities after disaster. And for the first time, emergency managers can use social intelligence to mitigate the harmful impacts of preexisting conditions prior to a disaster.

Finally, this book has been 5 years in the making. When we first started writing it, every emergency manager told us social intelligence (SI) is a worthless endeavor. Why? They believed that collecting SI would overwhelm emergency managers. Today, we are seeing community resiliency being discussed. The Federal Emergency Management Agency (FEMA) is beginning to plan for regional disasters in a more holistic fashion. Emergency managers are starting to account for disaster impact upon people and communities. More needs to be done. We hope you found social intelligence to be a useful tool in thinking about social vulnerability.

INDEX

9/11. *See* September 11 attacks

A

African Americans, poverty,
    perceptions of, 96–97. *See also*
    Poverty
Aid to Families with Dependent
    Children (AFDC), 94
Andrew, Hurricane
    evacuations, 23
    FEMA policies, impact of, 136
    government response, 24–25
    impact, 24
    National Guard involvement, 25,
      26–27
    overview, 23
    recovery, challenges to, 26–28
Art of War, 115

B

Baseline data points (BDPs), 129
    building departments, from, 200
    corporate, 141, 155–156
    GIS, use with, 175
    municipal, 140, 142–150
    nongovernmental entities (NGOs),
      140–141, 151–155
    nonprofit, 201–202
    overview, 139, 141
    public transit, 201
    school and education, 201
    social services, 201–202
Blair, Dennis C., 113
Blitzer, Wolf, 103
Bogart, William T., 194
Bond ratings, 8–9
BP Amoco, 132

BP oil spill. *See* Gulf Coast oil spill,
    2010
Brown, Michael, 92, 102
Bush, Barbara, 92
Bush, George, 98

C

Capital improvement planning, 198
Chambers of commerce, 184–185
Charities
    disaster recovery, role in, 57
Chicago
    budget deficit, 11
    evacuees taken in by, 11, 46–50
    heat wave of 1995, 58–59, 59–61,
      61–63, 63–65
    Illinois floods, 2010, 66, 69–71
    lessons learned, 210
    location quotient analysis (LQA),
      196–197
Children's Defense Fund, 106
Commonwealth Edison, 63
Community organizations active in
    disasters (COADs), 205–207
Coolidge, Calvin, 54
Culturalism, 96

D

Declarations of disasters, 89. *See also*
    Public policies, disasters
Deepwater Horizon oil spill. *See* Gulf
    Coast oil spill, 2010
Disaster policies. *See* Public policies,
    disasters
Disaster Relief Act, 1974, 88

217